SpringerBriefs in Optimization

Editors
Panos M. Pardalos
Industrial & Systems Engineering
University of Florida
Gainesville, Florida
USA

János D. Pintér
Pinter Consulting Services, Inc.
Halifax, Nova Scotia
Canada

Stephen Robinson
Industrial and Systems Engineering
University of Wisconsin
Madison, Wisconsin
USA

Tamás Terlaky
Industrial & Systems Engineering
Lehigh University
Bethlehem, Pennsylvania
USA

My T. Thai
Computer and Information Science and Engineering
University of Florida
Gainesville, Florida
USA

T0214188

SpringerBriefs in Optimization showcases algorithmic and theoretical techniques, case studies, and applications within the broad-based field of optimization. Manuscripts related to the ever-growing applications of optimization in applied mathematics, engineering, medicine, economics, and other applied sciences are encouraged.

More information about this series at http://www.springer.com/series/8918

Alexander J. Zaslavski

Stability of the Turnpike Phenomenon in Discrete-Time Optimal Control Problems

 Springer

Alexander J. Zaslavski
Department of Mathematics
Technion- Israel Institute of Techn
Haifa
Israel

ISSN 2190-8354   ISSN 2191-575X (electronic)
ISBN 978-3-319-08033-8   ISBN 978-3-319-08034-5 (eBook)
DOI 10.1007/978-3-319-08034-5
Springer Cham Heidelberg New York Dordrecht London

Library of Congress Control Number: 2014943649

Printed on acid-free paper

Springer is part of Springer Science+Business Media (www.springer.com)

Preface

The monograph is devoted to the study of the structure of approximate solutions of nonconvex (nonconcave) discrete-time optimal control problems. It contains new results on properties of approximate solutions which are independent of the length of the interval, for all sufficiently large intervals. These results deal with the so-called turnpike property of optimal control problems. The term was first coined by P. Samuelson in 1948 when he showed that an efficient expanding economy would spend most of the time in the vicinity of a balanced equilibrium path (also called a von Neumann path). To have the turnpike property means, roughly speaking, that the approximate solutions of the problems are determined mainly by the objective function and are essentially independent of the choice of interval and endpoint conditions, except in regions close to the endpoints. Now it is well-known that the turnpike property is a general phenomenon which holds for large classes of variational and optimal control problems. Using the Baire category (generic) approach, it was shown that the turnpike property holds for a generic (typical) variational problem [45] and for a generic optimal control problem [56]. According to the generic approach we say that a property holds for a generic (typical) element of a complete metric space (or the property holds generically) if the set of all elements of the metric space possessing this property contains a G? everywhere dense subset of the metric space which is a countable intersection of open everywhere dense sets. In [55] we were interested in individual (non-generic) turnpike results and in sufficient and necessary conditions for the turnpike phenomenon in the calculus of variations. In our recent research [46-51, 54] we were are also interested in individual turnpike results but for discrete-time optimal control problems which, in particular, describe a general model of economic dynamics. For these problems we established the turnpike property for approximate solutions with a singleton-turnpike and studied the stability of the turnpike phenomenon under small perturbations of objective functions.

In this book we continue to study the discrete-time optimal control problems considered in [46-51, 54]. Some results of these works are discussed in Chap. 1. In Chaps. 2 and 3 we show the stability of the turnpike phenomenon under small perturbations of objective functions and under small perturbations of control maps. The optimal control problems without discounting are studied in Chap. 2 while the discount case is considered in Chap. 3. In Chap. 4 we establish the turnpike property and its stability for discrete-time problems with nonsingleton-turnpikes. Note that the stability of the turnpike property is crucial in practice. One reason is that in practice we deal with a problem which consists of a perturbation of the problem we wish to consider. Another reason is that the computations introduce numerical errors.

Rishon LeZion   Alexander J. Zaslavski
December 30, 2013

Contents

List of Symbols

Chapter 1
Introduction

The study of the existence and the structure of solutions of optimal control problems defined on infinite intervals and on sufficiently large intervals has recently been a rapidly growing area of research. See, for example, [3, 4, 7–11, 13, 14, 16, 18–22, 25, 27, 32–34, 36, 44, 45, 52, 55] and the references mentioned therein. These problems arise in engineering [1, 23, 57], in models of economic growth [2, 5, 12, 13, 17, 22, 26, 31, 35, 38, 39, 40, 45], in infinite discrete models of solid-state physics related to dislocations in one-dimensional crystals [6, 41] and in the theory of thermodynamical equilibrium for materials [15, 24, 28–30]. In this chapter we discuss the structure of solutions of a discrete-time optimal control system describing a general model of economic dynamics.

1.1 The Turnpike Phenomenon

We study the structure of approximate solutions of an autonomous discrete-time control system with a compact metric space of states X equipped with a metric ρ. This control system is described by a bounded upper semicontinuous function $v : X \times X \to R^1$ which determines an optimality criterion and by a nonempty closed set $\Omega \subset X \times X$ which determines a class of admissible trajectories (programs). In models of economic growth the set X is the space of states, v is a utility function and $v(x_t, x_{t+1})$ evaluates consumption at moment t.

Consider the problem

$$\sum_{i=0}^{T-1} v(x_i, x_{i+1}) \to \max, \ \{(x_i, x_{i+1})\}_{i=0}^{T-1} \subset \Omega, \ x_0 = z, \ x_T = y, \qquad (1.1)$$

where $T \geq 1$ is an integer and the points $y, z \in X$.

We are interested in the turnpike property of approximate solutions which is independent of the length of the interval, for all sufficiently large intervals. To have this property means, roughly speaking, that approximate solutions of optimal control problems on an interval $[0, T]$ with given values y, z at the endpoints 0 and T,

A. J. Zaslavski, *Stability of the Turnpike Phenomenon in Discrete-Time Optimal Control Problems*, SpringerBriefs in Optimization, DOI 10.1007/978-3-319-08034-5_1, © The Author 2014

corresponding to the pair (v, Ω), are determined mainly by the objective function v, and are essentially independent of T, y and z.

In the classical turnpike theory the objective function v possesses the turnpike property (TP) if there exists a point $\bar{x} \in X$ (a turnpike) such that the following condition holds:

For each positive number ϵ there exists an integer $L \geq 1$ such that for each integer $T \geq 2L$ and each solution $\{x_i\}_{i=0}^{T} \subset X$ of the problem (P) the inequality $\rho(x_i, \bar{x}) \leq \epsilon$ is true for all $i = L, \ldots, T - L$.

It should be mentioned that the constant L depends neither on T nor on y, z.

The turnpike phenomenon has the following interpretation. If one wishes to reach a point A from a point B by a car in an optimal way, then one should turn to a turnpike, spend most of time on it and then leave the turnpike to reach the required point.

In the classical turnpike theory [17, 31, 38, 40] the space X is a compact convex subset of a finite-dimensional Euclidean space, the set Ω is convex and the function v is strictly concave. Under these assumptions the turnpike property can be established and the turnpike $\bar{x}$ is a unique solution of the maximization problem $v(x, x) \to$ max, $(x, x) \in \Omega$. In this situation it is shown that for each admissible sequence $\{x_t\}_{t=0}^{\infty}$ either the sequence $\{\sum_{t=0}^{T-1} v(x_t, x_{t+1}) - Tv(\bar{x}, \bar{x})\}_{T=1}^{\infty}$ is bounded (in this case the sequence $\{x_t\}_{t=0}^{\infty}$ is called (v)-good) or it diverges to $-\infty$. Moreover, it is also established that any (v)-good admissible sequence converges to the turnpike $\bar{x}$. In the sequel this property is called as the asymptotic turnpike property.

Recently it was shown that the turnpike property is a general phenomenon which holds for large classes of variational and optimal control problems without convexity assumptions. (See, for example, [45, 55, 56] and the references mentioned therein). For these classes of problems a turnpike is not necessarily a singleton but may instead be an nonstationary trajectory (in the discrete time nonautonomous case) or an absolutely continuous function on the interval $[0, \infty)$ (in the continuous time nonautonomous case) or a compact subset of the space X (in the autonomous case).

For classes of problems considered in [45, 56], using the Baire category approach, it was shown that the turnpike property holds for a generic (typical) problem. In this book we are interested in individual (non-generic) turnpike results and in stability of the turnpike phenomenon under small perturbations of the objective function v and the set Ω.

As we have mentioned before in general a turnpike is not necessarily a singleton. Nevertheless problems of the type (P) for which the turnpike is a singleton are of great importance because of the following reasons: there are many models of economic growth for which a turnpike is a singleton; if a turnpike is a singleton, then approximate solutions of (P) have very simple structure and this is very important for applications; if a turnpike is a singleton, then it can be easily calculated as a solution of the problem $v(x, x) \to$ max, $(x, x) \in \Omega$.

The turnpike property is very important for applications. Suppose that our objective function v has the turnpike property and we know a finite number of "approximate" solutions of the problem (P). Then we know the turnpike $\bar{x}$, or at least its approximation, and the constant L (see the definition of (TP)) which is an estimate for the time period required to reach the turnpike. This information can be useful if we need to find an "approximate" solution of the problem (P) with a new time interval $[m_1, m_2]$ and the new values $z, y \in X$ at the end points m_1 and m_2. Namely instead of solving this new problem on the "large" interval $[m_1, m_2]$ we can find an "approximate" solution of the problem (P) on the "small" interval $[m_1, m_1 + L]$ with the values $z, \bar{x}$ at the end points and an "approximate" solution of the problem (P) on the "small" interval $[m_2 - L, m_2]$ with the values $\bar{x}, y$ at the end points. Then the concatenation of the first solution, the constant sequence $x_i = \bar{x}$, $i = m_1 + L, \ldots, m_2 - L$ and the second solution is an "approximate" solution of the problem (P) on the interval $[m_1, m_2]$ with the values z, y at the end points. Sometimes as an "approximate" solution of the problem (P) we can choose any admissible sequence $\{x_i\}_{i=m_1}^{m_2}$ satisfying

$$x_{m_1} = z, \ x_{m_2} = y \ \text{and} \ x_i = \bar{x} \ \text{for all} \ i = m_1 + L, \ldots, m_2 - L.$$

1.2 Discrete-Time Problems

Let (X, ρ) be a compact metric space, Ω be a nonempty closed subset of $X \times X$ and let $v : X \times X \to R^1$ be a bounded upper semicontinuous function.

A sequence $\{x_t\}_{t=0}^{\infty} \subset X$ is called program if $(x_t, x_{t+1}) \in \Omega$ for all nonnegative integers t. A sequence $\{x_t\}_{t=0}^{T}$ where $T \geq 1$ is an integer is called a program if $(x_t, x_{t+1}) \in \Omega$ for all integers $t \in [0, T-1]$.

We consider the problems

$$\sum_{i=0}^{T-1} v(x_i, x_{i+1}) \to \max, \ \{(x_i, x_{i+1})\}_{i=0}^{T-1} \subset \Omega, \ x_0 = y,$$

and

$$\sum_{i=0}^{T-1} v(x_i, x_{i+1}) \to \max, \ \{(x_i, x_{i+1})\}_{i=0}^{T-1} \subset \Omega, \ x_0 = y, \ x_T = z,$$

where $T \geq 1$ is an integer and the points $y, z \in X$.

We suppose that there exist a point $\bar{x} \in X$ and a positive number $\bar{c}$ such that the following assumptions hold:

(i) $(\bar{x}, \bar{x})$ is an interior point of Ω and the function v is continuous at the point $(\bar{x}, \bar{x})$;

(ii) $\sum_{t=0}^{T-1} v(x_t, x_{t+1}) \leq Tv(\bar{x}, \bar{x}) + \bar{c}$ for any natural number T and any program $\{x_t\}_{t=0}^{T}$.

The property (ii) implies that for each program $\{x_t\}_{t=0}^{\infty}$ either the sequence

$$\left\{ \sum_{t=0}^{T-1} v(x_t, x_{t+1}) - Tv(\bar{x}, \bar{x}) \right\}_{T=1}^{\infty}$$

is bounded or $\lim_{T \to \infty} \left[\sum_{t=0}^{T-1} v(x_t, x_{t+1}) - Tv(\bar{x}, \bar{x}) \right] = -\infty.$

A program $\{x_t\}_{t=0}^{\infty}$ is called (v)-good if the sequence

$$\left\{ \sum_{t=0}^{T-1} v(x_t, x_{t+1}) - Tv(\bar{x}, \bar{x}) \right\}_{T=1}^{\infty}$$

is bounded.

Suppose that the following assumption holds.

(iii) (the asymptotic turnpike property) For any (v)-good program $\{x_t\}_{t=0}^{\infty}$, $\lim_{t\to\infty} \rho(x_t, \bar{x}) = 0$.

Note that the properties (i)–(iii) hold for models of economic dynamics considered in the classical turnpike theory.

For each positive number M denote by X_M the set of all points $x \in X$ for which there exists a program $\{x_t\}_{t=0}^{\infty}$ such that $x_0 = x$ and that for all natural numbers T the following inequality holds:

$$\sum_{t=0}^{T-1} v(x_t, x_{t+1}) - Tv(\bar{x}, \bar{x}) \geq -M.$$

It is not difficult to see that $\cup \{X_M : M \in (0, \infty)\}$ is the set of all points $x \in X$ for which there exists a (v)-good program $\{x_t\}_{t=0}^{\infty}$ satisfying $x_0 = x$. Let $u : X \times X \to R^1$, be a bounded function, an integer $T \geq 1$ and $\Delta \geq 0$. A program $\{x_i\}_{i=0}^{T} \subset X$ is called (u, Δ)-optimal if for any program $\{x_i'\}_{i=0}^{T}$ satisfying $x_0 = x_0'$, the inequality

$$\sum_{i=0}^{T-1} u(x_i, x_{i+1}) \geq \sum_{i=0}^{T-1} u(x_i', x_{i+1}') - \Delta$$

holds.

The following turnpike result describes the structure of approximate solutions of our first optimization problem stated above.

Theorem 1.1 *Let ϵ, M be positive numbers. Then there exist a natural number L and a positive number δ such that for each integer $T > 2L$ and each (v, δ)-optimal program $\{x_t\}_{t=0}^{T}$ which satisfies $x_0 \in X_M$ there exist nonnegative integers $\tau_1, \tau_2 \leq L$ such that $\rho(x_t, \bar{x}) \leq \epsilon$ for all $t = \tau_1, \ldots, T - \tau_2$ and if $\rho(x_0, \bar{x}) \leq \delta$, then $\tau_1 = 0$.*

An analogous turnpike result also holds for approximate solutions of our second optimization problem.

The following notion of an overtaking optimal program was introduced in [5, 17, 40].

A program $\{x_t\}_{t=0}^{\infty}$ is called (v)-overtaking optimal if for each program $\{y_t\}_{t=0}^{\infty}$ satisfying $y_0 = x_0$ the inequality

$$\limsup_{T\to\infty} \sum_{t=0}^{T-1} \left[v(y_t, y_{t+1}) - v(x_t, x_{t+1}) \right] \leq 0$$

holds.

The following result establishes the existence of an overtaking optimal program.

Theorem 1.2 *Assume that $x \in X$ and that there exists a (v)-good program $\{x_t\}_{t=0}^{\infty}$ such that $x_0 = x$. Then there exists a (v)-overtaking optimal program $\{x_t^*\}_{t=0}^{\infty}$ such that $x_0^* = x$.*

The following result provides necessary and sufficient conditions for overtaking optimality.

Theorem 1.3 *Let $\{x_t\}_{t=0}^{\infty}$ be a program such that*

$$x_0 \in \cup\{X_M : M \in (0, \infty)\}.$$

Then the program $\{x_t\}_{t=0}^{\infty}$ is (v)-overtaking optimal if and only if the following conditions hold:

(1) $\lim_{t \to \infty} \rho(x_t, \bar{x}) = 0$;
(2) *for each natural number T and each program $\{y_t\}_{t=0}^{T}$ satisfying $y_0 = x_0$, $y_T = x_T$ the inequality $\sum_{t=0}^{T-1} v(y_t, y_{t+1}) \leq \sum_{t=0}^{T-1} v(x_t, x_{t+1})$ holds.*

The next two theorems establish uniform convergence of overtaking optimal programs to $\bar{x}$.

Theorem 1.4 *Assume that the function v is continuous and let ϵ be a positive number. Then there exists a positive number δ such that for each (v)-overtaking optimal program $\{x_t\}_{t=0}^{\infty}$ satisfying $\rho(x_0, \bar{x}) \leq \delta$ the inequality $\rho(x_t, \bar{x}) \leq \epsilon$ holds for all nonnegative integers t.*

Theorem 1.5 *Assume that the function v is continuous and let M, ϵ be positive numbers. Then there exists an integer $L \geq 1$ such that for each (v)-overtaking optimal program $\{x_t\}_{t=0}^{\infty}$ satisfying $x_0 \in X_M$ the inequality $\rho(x_t, \bar{x}) \leq \epsilon$ holds for all integers $t \geq L$.*

Theorems 1.1–1.5 were obtained in [46].

Example 1.6 Let (X, ρ) be a compact metric space, Ω be a nonempty closed subset of $X \times X$, $\bar{x} \in X$, $(\bar{x}, \bar{x})$ be an interior point of Ω, $\pi : X \to R^1$ be a continuous function, α be a real number and $L : X \times X \to [0, \infty)$ be a continuous function such that for each $(x, y) \in X \times X$ the equality $L(x, y) = 0$ holds if and only if $(x, y) = (\bar{x}, \bar{x})$. Set

$$v(x, y) = \alpha - L(x, y) + \pi(x) - \pi(y)$$

for all $x, y \in X$. It is not difficult to see that assumptions (i), (ii) and (iii) hold.

Example 1.7 Let X be a compact convex subset of the Euclidean space R^n with the norm $|\cdot|$ induced by the scalar product $\langle \cdot, \cdot \rangle$, let $\rho(x, y) = |x - y|$, $x, y \in R^n$, Ω be a nonempty closed subset of $X \times X$, a point $\bar{x} \in X$, $(\bar{x}, \bar{x})$ be an interior point of Ω and let $v : X \times X \to R^1$ be a strictly concave continuous function such that

$$v(\bar{x}, \bar{x}) = \sup\{v(z, z) : z \in X \text{ and } (z, z) \in \Omega\}.$$

We assume that there exists a positive constant $\bar{r}$ such that

$$\{(x, y) \in R^n \times R^n : |x - \bar{x}|, |y - \bar{x}| \leq \bar{r}\} \subset \Omega.$$

It is a well-known fact of convex analysis [37] that there exists a point $l \in R^n$ such that

$$v(x, y) \le v(\bar{x}, \bar{x}) + \langle l, x - y \rangle$$

for any point $(x, y) \in X \times X$. Set

$$L(x, y) = v(\bar{x}, \bar{x}) + \langle l, x - y \rangle - v(x, y)$$

for all $(x, y) \in X \times X$. It is not difficult to see that this example is a particular case of Example 1.6. Therefore assumptions (i), (ii) and (iii) hold.

Denote by $\mathcal{M}$ the set of all bounded functions $u : X \times X \to R^1$. For each function $w \in \mathcal{M}$ we set

$$\|w\| = \sup\{|w(x, y)| : x, y \in X\}.$$

The following two theorems obtained in [48, 49] respectively show that the turnpike property is stable under perturbations of the objective function.

Theorem 1.8 *Let $M_0, \epsilon > 0$. Then there exist a positive number δ and an integer $L_* \ge 1$ such that for each function $u \in \mathcal{M}$ satisfying $\|u - v\| \le \delta$, each integer $T > 2L_*$ and each (u, δ)-optimal program $\{x_t\}_{t=0}^T$ which satisfies $x_0 \in X_{M_0}$ there exist integers $\tau_1 \in [0, L_*], \tau_2 \in [T - L_*, T]$ such that*

$$\rho(x_t, \bar{x}) \le \epsilon, \ t = \tau_1, \dots, \tau_2.$$

Moreover if $\rho(x_0, \bar{x}) \le \delta$, then $\tau_1 = 0$.
Denote by $\mathrm{Card}(A)$ the cardinality of a set A.

Theorem 1.9 *Let M_0, M_1, ϵ be positive numbers. Then there exist a positive number δ and an integer $L_* \ge 1$ such that for each function $u \in \mathcal{M}$ satisfying $\|u - v\| \le \delta$, each integer $T > L_*$ and each (u, M_1)-optimal program $\{x_t\}_{t=0}^T$ which satisfies $x_0 \in X_{M_0}$ the following inequality holds:*

$$\mathrm{Card}(\{t \in \{0, \dots, T\} : \rho(x_t, \bar{x}) > \epsilon\}) \le L_*.$$

1.3 Examples

Example 1.10 Let $X = [0, 1], \Omega = \{(x, y) \in [0, 1] \times [0, 1] : y \le x^{1/2}\}, v(x, y) = x^{1/2} - y^2, x, y \in X$. It is not difficult to see that the set Ω is convex, the function v is strictly concave, the optimization problem $v(z, z) \to \max, z \in X$ and $(z, z) \in \Omega$ has a unique solution $16^{-1/3}$ and $\left(16^{-1/3}, 16^{-1/3}\right)$ is an interior point of Ω. Therefore this example is a particular case of Example 1.7 and assumptions (i), (ii) and (iii) hold.

Example 1.11 Let $X = [0, 1], \Omega = \{(x, y) \in [0, 1] \times [0, 1] : y \le x^{1/2}\}, v(x, y) = x^{1/2} - y, x, y \in X$. It is not difficult to see that the set Ω is convex, the function v is concave but not strictly concave, the optimization problem $v(z, z) \to \max, z \in X$

and $(z, z) \in \Omega$ has a unique solution 4^{-1} and $(4^{-1}, 4^{-1})$ is an interior point of Ω. Since the function v is concave for all $x, y \in X$,

$$v(x, y) \leq v(4^{-1}, 4^{-1}) + x - y = 4^{-1} + x - y$$

and

$$4^{-1} + x - y - v(x, y) = \left(x^{1/2} - 2^{-1}\right)^2$$

is equal zero if and only if $x = 4^{-1}$. Now it is not difficult to see that assumptions (i), (ii) and (iii) hold.

Example 1.12 Consider the sets X, Ω and the function v defined in Example 1.11 and set $u(x, y) = x^{1/2} - x^2 - y + y^2$, $x, y \in X$. The function u is strictly convex with respect to the variable y. Nevertheless assumptions (i), (ii) and (iii) hold for the function u because for any integer T and any program $\{x_t\}_{t=0}^T$,

$$\sum_{t=0}^{T-1} u(x_t, x_{t+1}) = \sum_{t=0}^{T-1} v(x_t, x_{t+1}) + x_T^2 - x_0^2.$$

Chapter 2
Optimal Control Problems with Singleton Turnpikes

In this chapter we study the structure of solutions of a discrete-time control system with a compact metric space of states X which arises in economic dynamics. This control system is described by a nonempty closed set $\Omega \subset X \times X$ which determines a class of admissible trajectories (programs) and by a bounded upper semicontinuous objective function $v : X \times X \to R^1$ which determines an optimality criterion. We show the stability of the turnpike phenomenon under small perturbations of the objective function v and the set Ω.

2.1 Preliminaries and Stability Results

Let (X, ρ) be a compact metric space. For each $x \in X$ and each nonempty set $C \subset X$ set

$$\rho(x, C) = \inf\{\rho(x, y) : y \in C\}.$$

For each $x \in X$ and each $r > 0$ set

$$B(x, r) = \{y \in X : \rho(x, y) \le r\}.$$

We equip the space $X \times X$ with the metric ρ_1 defined by

$$\rho_1((x_1, x_2), (y_1, y_2)) = \rho(x_1, y_1) + \rho(x_2, y_2), \quad x_1, x_2, y_1, y_2 \in X.$$

For each $(x_1, x_2) \in X \times X$ and each nonempty set $C \subset X \times X$ set

$$\rho_1((x_1, x_2), C) = \inf\{\rho_1((x_1, x_2), (y_1, y_2)) : (y_1, y_2) \in C\}.$$

Denote by $\mathcal{M}$ the set of all bounded functions $u : X \times X \to R^1$. For each $w \in \mathcal{M}$ set

$$\|w\| = \sup\{|w(x, y)| : (x, y) \in X \times X\}.$$

Let Ω be a nonempty closed subset of $X \times X$.

A. J. Zaslavski, *Stability of the Turnpike Phenomenon in Discrete-Time Optimal Control Problems*, SpringerBriefs in Optimization, DOI 10.1007/978-3-319-08034-5_2, © The Author 2014

A sequence $\{x_t\}_{t=0}^{\infty} \subset X$ is called an (Ω)-program if $(x_t, x_{t+1}) \in \Omega$ for all integers $t \geq 0$. A sequence $\{x_t\}_{t=T_1}^{T_2} \subset X$ where integers T_1, T_2 satisfy $0 \leq T_1 < T_2$ is called an (Ω)-program if $(x_t, x_{t+1}) \in \Omega$ for all integers $t \in [T_1, T_2 - 1]$.

Let $v \in \mathcal{M}$ be an upper semicontinuous function.

We suppose that there exist $\bar{x} \in X$ and a constant $\bar{c} > 0$ such that the following assumptions hold.

(A1) $(\bar{x}, \bar{x})$ is an interior point of Ω (there is $\epsilon > 0$ such that $\{(x, y) \in X \times X : \rho(x, \bar{x}), \ \rho(y, \bar{x}) \leq \epsilon\} \subset \Omega$) and v is continuous at $(\bar{x}, \bar{x})$.

(A2) For any integer $T \geq 1$ and any (Ω)-program $\{x_t\}_{t=0}^{T}$,

$$\sum_{t=0}^{T-1} v(x_t, x_{t+1}) \leq Tv(\bar{x}, \bar{x}) + \bar{c}.$$

Assumption (A2) implies the following result.

Proposition 2.1 *For each (Ω)-program $\{x_t\}_{t=0}^{\infty}$ either the sequence*

$$\left\{\sum_{t=0}^{T-1} v(x_t, x_{t+1}) - Tv(\bar{x}, \bar{x})\right\}_{T=1}^{\infty}$$

is bounded or $\lim_{T \to \infty}\left[\sum_{t=0}^{T-1} v(x_t, x_{t+1}) - Tv(\bar{x}, \bar{x})\right] = -\infty.$

An (Ω)-program $\{x_t\}_{t=0}^{\infty}$ is called (v, Ω)-good if the sequence

$$\left\{\sum_{t=0}^{T-1} v(x_t, x_{t+1}) - Tv(\bar{x}, \bar{x})\right\}_{T=1}^{\infty}$$

is bounded [13, 17, 45, 55, 56].

In this chapter we suppose that the following assumption holds.

(A3) (the asymptotic turnpike property) For any (v, Ω)-good program $\{x_t\}_{t=0}^{\infty}$, $\lim_{t \to \infty} \rho(x_t, \bar{x}) = 0$.

Note that (A3) holds for many important infinite horizon optimal control problems. In particular, (A3) holds for a general model of economic dynamics considered in Example 1.7.

For each $x, y \in X$, each integer $T \geq 1$ and each $w \in \mathcal{M}$ set

$$\sigma(w, T, x, y)$$

$$= \sup\left\{\sum_{i=0}^{T-1} w(x_i, x_{i+1}) : \{x_i\}_{i=0}^{T} \text{ is an } (\Omega) - \text{program and } x_0 = x, \ x_T = y\right\}.$$

(Here we use the convention that the supremum of an empty set is $-\infty$.)

In Chap. 1 we considered the turnpike properties of approximate solutions of the problems

$$\sum_{i=0}^{T-1} v(x_i, x_{i+1}) \to \max, \ \{(x_i, x_{i+1})\}_{i=0}^{T-1} \subset \Omega, \ x_0 = y,$$

and

$$\sum_{i=0}^{T-1} v(x_i, x_{i+1}) \to \max, \ \{(x_i, x_{i+1})\}_{i=0}^{T-1} \subset \Omega, \ x_0 = y, \ x_T = z,$$

where $T \geq 1$ is an integer and the points $y, z \in X$.

In this chapter we show that these turnpike properties are stable under small perturbations of the objective function v and the set Ω. In order to meet this goal we introduce the following definitions.

By assumption (A1) there exists $\bar{r} \in (0, 1)$ such that

$$B(\bar{x}, \bar{r}) \times B(\bar{x}, \bar{r}) \subset \Omega. \tag{2.1}$$

Fix

$$\bar{\lambda} \in (0, \bar{r}). \tag{2.2}$$

For each $\lambda > 0$ denote by $\mathcal{E}(\lambda)$ the collection of all nonempty sets $\Omega' \subset X \times X$ such that

$$\rho_1(z, \Omega) \leq \lambda \text{ for each } z \in \Omega', \tag{2.3}$$

$$B(\bar{x}, \bar{\lambda}) \times B(\bar{x}, \bar{\lambda}) \subset \Omega'. \tag{2.4}$$

Let integers T_1, T_2 satisfy $0 \leq T_1 < T_2$ and let $\Omega_t, t = T_1, \ldots, T_2 - 1$ be nonempty subsets of $X \times X$.

A sequence $\{x_t\}_{t=T_1}^{T_2} \subset X$ is called an $(\{\Omega_t\}_{t=T_1}^{T_2-1})$-program if $(x_t, x_{t+1}) \in \Omega_t$ for all integers $t \in [T_1, T_2 - 1]$.

For each $x, y \in X$ and each finite sequence $\{u_t\}_{t=T_1}^{T_2-1} \subset \mathcal{M}$ set

$$\sigma(\{u_t\}_{t=T_1}^{T_2-1}, \{\Omega_t\}_{t=T_1}^{T_2-1}, T_1, T_2, x) = \sup \left\{ \sum_{t=T_1}^{T_2-1} u_t(x_t, x_{t+1}) : \right.$$

$$\left. \{x_t\}_{t=T_1}^{T_2} \text{ is an } (\{\Omega_t\}_{t=T_1}^{T_2-1}) - \text{program and } x_{T_1} = x \right\}, \tag{2.5}$$

$$\sigma\left(\{u_t\}_{t=T_1}^{T_2-1}, \{\Omega_t\}_{t=T_1}^{T_2-1}, T_1, T_2, x, y\right) = \sup \left\{ \sum_{t=T_1}^{T_2-1} u_t(x_t, x_{t+1}) : \right.$$

$$\left. \{x_t\}_{t=T_1}^{T_2} \text{ is an } (\{\Omega_t\}_{t=T_1}^{T_2-1}) - \text{program}, \ x_{T_1} = x \text{ and } x_{T_2} = y \right\}, \tag{2.6}$$

$$\sigma(\{u_t\}_{t=T_1}^{T_2-1}, \{\Omega_t\}_{t=T_1}^{T_2-1}, T_1, T_2) = \sup\left\{\sum_{t=T_1}^{T_2-1} u_t(x_t, x_{t+1}):\right.$$

$$\left.\{x_t\}_{t=T_1}^{T_2} \text{ is an } (\{\Omega_t\}_{t=T_1}^{T_2-1}) - \text{program}\right\}. \qquad (2.7)$$

(Here we use the convention that the supremum of an empty set is $-\infty$).

Denote by $Y(\{\Omega_t\}_{t=T_1}^{T_2-1}, T_1, T_2)$ the set of all $x \in X$ for which there exists an $(\{\Omega_t\}_{t=T_1}^{T_2-1})$-program $\{x_t\}_{t=T_1}^{T_2}$ such that $x_{T_1} = \bar{x}$ and $x_{T_2} = x$.

Denote by $\bar{Y}(\{\Omega_t\}_{t=T_1}^{T_2-1}, T_1, T_2)$ the set of all $x \in X$ for which there exists an $(\{\Omega_t\}_{t=T_1}^{T_2-1})$-program $\{x_t\}_{t=T_1}^{T_2}$ such that $x_{T_1} = x$ and $x_{T_2} = \bar{x}$.

For sufficiently small positive numbers δ, we study the structure of approximate solutions of the problems

$$\sum_{i=0}^{T-1} u_i(x_i, x_{i+1}) \to \max,$$

$$\{x_i\}_{i=0}^{T} \text{ is an } (\{\Omega_t\}_{t=0}^{T-1}) - \text{program and } x_0 = y,$$

and

$$\sum_{i=0}^{T-1} u_i(x_i, x_{i+1}) \to \max,$$

$$\{x_i\}_{i=0}^{T} \text{ is an } (\{\Omega_t\}_{t=0}^{T-1}) - \text{program and } x_0 = y, \ x_T = z,$$

where $T \geq 1$ is an integer, $y, z \in X$ and for all $t = 0, \ldots, T-1$, we have

$$\Omega_t \in \mathcal{E}(\delta), \ u_t \in \mathcal{M} \text{ and } \|u_t - v\| \leq \delta.$$

In this chapter we prove the following four stability results.

Theorem 2.2 *Let ϵ be a positive number and let l_1, l_2 be natural numbers. Then there exist $\delta > 0$ and a natural number $L > l_1 + l_2$ such that for each integer $T > 2L$, each*

$$\Omega_t \in \mathcal{E}(\delta), \ t = 0, \ldots, T-1,$$

each $u_t \in \mathcal{M}$, $t = 0, \ldots, T-1$ satisfying

$$\|u_t - v\| \leq \delta, \ t = 0, \ldots, T-1$$

and each $(\{\Omega_t\}_{t=0}^{T-1})$-program $\{x_t\}_{t=0}^{T}$ which satisfies

$$x_0 \in \bar{Y}(\{\Omega_t\}_{t=0}^{l_1-1}, 0, l_1), \ x_T \in Y(\{\Omega_t\}_{t=T-l_2}^{T-1}, T-l_2, T),$$

$$\sigma(\{u_t\}_{t=0}^{T-1}, \{\Omega_t\}_{t=0}^{T-1}, 0, T, x_0, x_T) \leq \sum_{t=0}^{T-1} u_t(x_t, x_{t+1}) + \delta$$

there exist integers $\tau_1 \in [0, L]$, $\tau_2 \in [T - L, T]$ such that

$$\rho(x_t, \bar{x}) \le \epsilon \text{ for all } t = \tau_1, \ldots, \tau_2.$$

Moreover if $\rho(x_0, \bar{x}) \le \delta$, then $\tau_1 = 0$ and if $\rho(x_T, \bar{x}) \le \delta$, then $\tau_2 = T$.

Theorem 2.3 *Let ϵ be a positive number and let l_1 be a natural number. Then there exist $\delta > 0$ and a natural number $L > l_1$ such that for each integer $T > 2L$, each*

$$\Omega_t \in \mathcal{E}(\delta), \ t = 0, \ldots, T - 1,$$

each $u_t \in \mathcal{M}$, $t = 0, \ldots, T - 1$ satisfying

$$\|u_t - v\| \le \delta, \ t = 0, \ldots, T - 1$$

and each $(\{\Omega_t\}_{t=0}^{T-1})$-program $\{x_t\}_{t=0}^{T}$ which satisfies

$$x_0 \in \bar{Y}(\{\Omega_t\}_{t=0}^{l_1-1}, 0, l_1),$$

$$\sigma(\{u_t\}_{t=0}^{T-1}, \{\Omega_t\}_{t=0}^{T-1}, 0, T, x_0) \le \sum_{t=0}^{T-1} u_t(x_t, x_{t+1}) + \delta$$

there exist integers $\tau_1 \in [0, L]$, $\tau_2 \in [T - L, T]$ such that

$$\rho(x_t, \bar{x}) \le \epsilon \text{ for all } t = \tau_1, \ldots, \tau_2.$$

Moreover if $\rho(x_0, \bar{x}) \le \delta$, then $\tau_1 = 0$ and if $\rho(x_T, \bar{x}) \le \delta$, then $\tau_2 = T$.
Denote by $\mathrm{Card}(B)$ the cardinality of a set B.

Theorem 2.4 *Let ϵ, M be positive numbers and let l_1, l_2 be natural numbers. Then there exist $\delta > 0$ and a natural number $L > l_1 + l_2$ such that for each integer $T > L$, each*

$$\Omega_t \in \mathcal{E}(\delta), \ t = 0, \ldots, T - 1,$$

each $u_t \in \mathcal{M}$, $t = 0, \ldots, T - 1$ satisfying

$$\|u_t - v\| \le \delta, \ t = 0, \ldots, T - 1$$

and each $(\{\Omega_t\}_{t=0}^{T-1})$-program $\{x_t\}_{t=0}^{T}$ which satisfies

$$x_0 \in \bar{Y}(\{\Omega_t\}_{t=0}^{l_1-1}, 0, l_1), \ x_T \in Y(\{\Omega_t\}_{t=T-l_2}^{T-1}, T - l_2, T),$$

$$\sigma(\{u_t\}_{t=0}^{T-1}, \{\Omega_t\}_{t=0}^{T-1}, 0, T, x_0, x_T) \le \sum_{t=0}^{T-1} u_t(x_t, x_{t+1}) + M$$

the inequality

$$\mathrm{Card}(\{t \in \{0, \ldots, T\} : \ \rho(x_t, \bar{x}) > \epsilon\}) \le L$$

holds.

Theorem 2.5 *Let ϵ, M be positive numbers and let l_1 be a natural number. Then there exist $\delta > 0$ and a natural number $L > l_1$ such that for each integer $T > L$, each*

$$\Omega_t \in \mathcal{E}(\delta), \ t = 0, \ldots, T - 1,$$

each $u_t \in \mathcal{M}, \ t = 0, \ldots, T - 1$ satisfying

$$\|u_t - v\| \leq \delta, \ t = 0, \ldots, T - 1$$

and each $(\{\Omega_t\}_{t=0}^{T-1})$-program $\{x_t\}_{t=0}^T$ which satisfies

$$x_0 \in \bar{Y}(\{\Omega_t\}_{t=0}^{l_1-1}, 0, l_1),$$

$$\sigma(\{u_t\}_{t=0}^{T-1}, \{\Omega_t\}_{t=0}^{T-1}, 0, T, x_0) \leq \sum_{t=0}^{T-1} u_t(x_t, x_{t+1}) + M$$

the inequality

$$Card(\{t \in \{0, \ldots, T\} : \ \rho(x_t, \bar{x}) > \epsilon\}) \leq L$$

holds.

2.2 Extensions

We use the notation, definitions, and assumptions introduced in Sect. 2.1. In this section we state the extensions of the turnpike results of the previous section. In these extensions we describe the structure of programs defined on an interval $[0, T]$ with sufficiently large T which are approximate solutions of the corresponding optimal problems on subintervals of the length L, where L is a constant which does not depend on T.

Theorem 2.6 *Let $\epsilon \in (0, \bar{\lambda})$ and M be a positive number. Then there exist $\gamma \in (0, \epsilon)$ and a natural number L_0 such that for each integer $L_1 \geq L_0$ there exists a positive number $\delta < \gamma$ such that the following assertion holds.*
    Assume that an integer $T > 3L_1$,

$$\Omega_t \in \mathcal{E}(\delta), \ t = 0, \ldots, T - 1,$$

$u_t \in \mathcal{M}, \ t = 0, \ldots, T - 1$ satisfy

$$\|u_t - v\| \leq \delta, \ t = 0, \ldots, T - 1$$

and that an $(\{\Omega_t\}_{t=0}^{T-1})$-program $\{x_t\}_{t=0}^T$ and a finite sequence of integers $\{S_i\}_{i=0}^q$ satisfy

$$S_0 = 0, \ S_{i+1} - S_i \in [L_0, L_1], \ i = 0, \ldots, q - 1, \ S_q > T - L_1,$$

$$\sum_{t=S_i}^{S_{i+1}-1} u_t(x_t, x_{t+1}) \geq \sum_{t=S_i}^{S_{i+1}-1} u_t(\bar{x}, \bar{x}) - M$$

for each integer $i \in [0, q-1]$,

$$\sum_{t=S_i}^{S_{i+2}-1} u_t(x_t, x_{t+1}) \geq \sigma(\{u_t\}_{t=S_i}^{S_{i+2}-1}, \{\Omega_t\}_{t=S_i}^{S_{i+2}-1}, S_i, S_{i+2}, x_{S_i}, x_{S_{i+2}}) - \gamma$$

for each integer $i \in [0, q-2]$ and

$$\sum_{t=S_{q-2}}^{T-1} u_t(x_t, x_{t+1}) \geq \sigma(\{u_t\}_{t=S_{q-2}}^{T-1}, \{\Omega_t\}_{t=S_{q-2}}^{T-1}, S_{q-2}, T, x_{S_{q-2}}, x_T) - \gamma.$$

Then there exist integers $\tau_1 \in [0, L_1]$, $\tau_2 \in [T - 2L_1, T]$ such that

$$\rho(x_t, \bar{x}) \leq \epsilon \text{ for all } t = \tau_1, \dots, \tau_2.$$

Moreover if $\rho(x_0, \bar{x}) \leq \gamma$, then $\tau_1 = 0$ and if $\rho(x_T, \bar{x}) \leq \gamma$, then $\tau_2 = T$.

Theorem 2.7 Let ϵ, M be positive numbers and let l_1, l_2 be natural numbers. Then there exist $\delta > 0$ and a natural number $L > l_1 + l_2$ such that for each integer $T > 2L$, each

$$\Omega_t \in \mathcal{E}(\delta), \ t = 0, \dots, T - 1,$$

each $u_t \in \mathcal{M}$, $t = 0, \dots, T - 1$ satisfying

$$\|u_t - v\| \leq \delta, \ t = 0, \dots, T - 1$$

and each $(\{\Omega_t\}_{t=0}^{T-1})$-program $\{x_t\}_{t=0}^{T}$ which satisfies

$$x_0 \in \bar{Y}(\{\Omega_t\}_{t=0}^{l_1-1}, 0, l_1), \ x_T \in Y(\{\Omega_t\}_{t=T-l_2}^{T-1}, T - l_2, T),$$

$$\sigma(\{u_t\}_{t=0}^{T-1}, \{\Omega_t\}_{t=0}^{T-1}, 0, T, x_0, x_T) \leq \sum_{t=0}^{T-1} u_t(x_t, x_{t+1}) + M$$

and

$$\sum_{t=\tau}^{\tau+L-1} u_t(x_t, x_{t+1}) \geq \sigma(\{u_t\}_{t=\tau}^{\tau+L-1}, \{\Omega_t\}_{t=\tau}^{\tau+L-1}, \tau, \tau + L, x_\tau, x_{\tau+L}) - \delta$$

for each integer $\tau \in [0, T - L]$, there exist integers $\tau_1 \in [0, L]$, $\tau_2 \in [T - L, T]$ such that

$$\rho(x_t, \bar{x}) \leq \epsilon \text{ for all } t = \tau_1, \dots, \tau_2.$$

Moreover if $\rho(x_0, \bar{x}) \leq \delta$, then $\tau_1 = 0$ and if $\rho(x_T, \bar{x}) \leq \delta$, then $\tau_2 = T$.

Theorem 2.8 *Let ϵ, M be positive numbers and let l_1 be a natural number. Then there exist $\delta > 0$ and a natural number $L > l_1$ such that for each integer $T > 2L$, each*

$$\Omega_t \in \mathcal{E}(\delta), \ t = 0, \ldots, T - 1,$$

each $u_t \in \mathcal{M}$, $t = 0, \ldots, T - 1$ satisfying

$$\|u_t - v\| \leq \delta, \ t = 0, \ldots, T - 1$$

and each $(\{\Omega_t\}_{t=0}^{T-1})$-program $\{x_t\}_{t=0}^{T}$ which satisfies

$$x_0 \in \bar{Y}(\{\Omega_t\}_{t=0}^{l_1-1}, 0, l_1),$$

$$\sigma(\{u_t\}_{t=0}^{T-1}, \{\Omega_t\}_{t=0}^{T-1}, 0, T, x_0) \leq \sum_{t=0}^{T-1} u_t(x_t, x_{t+1}) + M$$

and

$$\sum_{t=\tau}^{\tau+L-1} u_t(x_t, x_{t+1}) \geq \sigma(\{u_t\}_{t=\tau}^{\tau+L-1}, \{\Omega_t\}_{t=\tau}^{\tau+L-1}, \tau, \tau + L, x_\tau, x_{\tau+L}) - \delta$$

for each integer $\tau \in [0, T - L]$, there exist integers $\tau_1 \in [0, L]$, $\tau_2 \in [T - L, T]$ such that

$$\rho(x_t, \bar{x}) \leq \epsilon \ \text{for all} \ t = \tau_1, \ldots, \tau_2.$$

Moreover if $\rho(x_0, \bar{x}) \leq \delta$, then $\tau_1 = 0$ and if $\rho(x_T, \bar{x}) \leq \delta$, then $\tau_2 = T$.

2.3   Three Lemmata

In order to prove our stability results we need the following useful lemmas. Lemmas 2.9 and 2.10 were obtained in [46] while Lemma 2.11 was proved in [47].

Lemma 2.9 *Let $\epsilon > 0$ and $M_0 > 0$. Then there exists a natural number T such that for each (Ω)-program $\{x_t\}_{t=0}^{T}$ which satisfies*

$$\sum_{t=0}^{T-1} v(x_t, x_{t+1}) \geq Tv(\bar{x}, \bar{x}) - M_0$$

the relation

$$\min\{\rho(x_i, \bar{x}) : \ i = 1, \ldots, T\} \leq \epsilon$$

holds.

Proof Let us assume the contrary. Then for each natural number k there exists an (Ω)-program $\{x_t^{(k)}\}_{t=0}^{k}$ which satisfies

$$\sum_{t=0}^{k-1} v(x_t^{(k)}, x_{t+1}^{(k)}) \geq kv(\bar{x}, \bar{x}) - M_0, \tag{2.8}$$

$$\rho(x_t^{(k)}, \bar{x}) > \epsilon \text{ for all integers } t = 1, \ldots, k. \tag{2.9}$$

Let $k \geq 1$ be an integer. By (2.8) and (A2) for each integer j satisfying $0 < j < k$

$$\sum_{t=0}^{j-1} v(x_t^{(k)}, x_{t+1}^{(k)}) = \sum_{t=0}^{k-1} v(x_t^{(k)}, x_{t+1}^{(k)}) - \sum_{t=j}^{k-1} v(x_t^{(k)}, x_{t+1}^{(k)})$$

$$\geq kv(\bar{x}, \bar{x}) - M_0 - \sum_{t=j}^{k-1} v(x_t^{(k)}, x_{t+1}^{(k)})$$

$$\geq kv(\bar{x}, \bar{x}) - M_0 - (k - j)v(\bar{x}, \bar{x}) - \bar{c}.$$

Together with (2.8) this inequality implies that for each integer $k \geq 1$ and each $j \in \{1, \ldots, k\}$

$$\sum_{t=0}^{j-1} v(x_t^{(k)}, x_{t+1}^{(k)}) \geq jv(\bar{x}, \bar{x}) - \bar{c} - M_0. \tag{2.10}$$

There exists a strictly increasing sequence of natural numbers $\{k_i\}_{i=1}^{\infty}$ such that for each integer $t \geq 0$ there exists

$$x_t = \lim_{i \to \infty} x_t^{(k_i)}. \tag{2.11}$$

Clearly, $\{x_t\}_{t=0}^{\infty}$ is an (Ω)-program. In view of (2.11) and (2.9)

$$\rho(x_t, \bar{x}) \geq \epsilon \text{ for all integers } t \geq 1. \tag{2.12}$$

It follows from (2.11) and (2.10) that for each integer $T \geq 1$ we have

$$\sum_{t=0}^{T-1} v(x_t, x_{t+1}) \geq Tv(\bar{x}, \bar{x}) - M_0 - \bar{c}.$$

This implies that $\{x_t\}_{t=0}^{\infty}$ is a (v, Ω)-good program. By (A3) $\lim_{t \to \infty} \rho(x_t, \bar{x}) = 0$. This equality contradicts (2.12). The contradiction we have reached proves Lemma 2.9.

Lemma 2.10 Let $\epsilon > 0$. Then there exists $\delta > 0$ such that for each integer $T \geq 1$ and each (Ω)-program $\{x_t\}_{t=0}^{T}$ which satisfies

$$\rho(x_0, \bar{x}), \ \rho(x_T, \bar{x}) \leq \delta, \tag{2.13}$$

$$\sum_{t=0}^{T-1} v(x_t, x_{t+1}) \geq \sigma(v, T, x_0, x_T) - \delta \tag{2.14}$$

the inequality $\rho(x_t, \bar{x}) \leq \epsilon$ holds for all $t = 0, \ldots, T$.

Proof Since v is continuous at $(\bar{x}, \bar{x})$ for each natural number k there exists

$$\delta_k \in (0, 2^{-k}\bar{r}) \tag{2.15}$$

such that

$$|v(x, y) - v(\bar{x}, \bar{x})| \le 2^{-k} \tag{2.16}$$

for each $x, y \in X$ satisfying

$$\rho(x, \bar{x}), \ \rho(y, \bar{x}) \le \delta_k. \tag{2.17}$$

Assume that the lemma is wrong. Then for each natural number k there exist an integer $T_k \ge 1$ and an (Ω)-program $\{x_t^{(k)}\}_{t=0}^{T_k}$ such that

$$\rho\left(x_0^{(k)}, \bar{x}\right), \ \rho\left(x_{T_k}^{(k)}, \bar{x}\right) \le \delta_k, \tag{2.18}$$

$$\sum_{t=0}^{T_k-1} v\left(x_t^{(k)}, x_{t+1}^{(k)}\right) \ge \sigma\left(v, T_k, x_0^{(k)}, x_{T_k}^{(k)}\right) - \delta_k, \tag{2.19}$$

$$\max\{\rho\left(x_t^{(k)}, \bar{x}\right): \ t = 0, \ldots, T_k\} > \epsilon. \tag{2.20}$$

Let $k \ge 1$ be an integer. Define a sequence $\{z_t\}_{t=0}^{T_k} \subset X$ as follows:

$$z_0 = x_0^{(k)}, \ z_{T_k} = x_{T_k}^{(k)}, \ z_t = \bar{x}, \ t \in \{0, \ldots, T_k\} \setminus \{0, T_k\}. \tag{2.21}$$

By (2.21), (2.18), (2.15), and (2.1), $\{z_t\}_{t=0}^{T_k}$ is an (Ω)-program. It follows from (2.19) and (2.21) that

$$\sum_{t=0}^{T_k-1} v\left(x_t^{(k)}, x_{t+1}^{(k)}\right) \ge \sigma\left(v, T_k, x_0^{(k)}, x_{T_k}^{(k)}\right) - \delta_k \ge \sum_{t=0}^{T_k-1} v(z_t, z_{t+1}) - \delta_k. \tag{2.22}$$

In view of (2.18), (2.21), and the choice of δ_k (see (2.15)–(2.17))

$$|v(z_0, z_1) - v(\bar{x}, \bar{x})| \le 2^{-k}, |v(z_{T_k-1}, z_{T_k}) - v(\bar{x}, \bar{x})| \le 2^{-k},$$
$$v(z_t, z_{t+1}) = v(\bar{x}, \bar{x}), \ t \in \{0, \ldots, T_k-1\} \setminus \{0, T_k - 1\}. \tag{2.23}$$

Relations (2.23) and (2.22) imply that

$$\sum_{t=0}^{T_k-1} v\left(x_t^{(k)}, x_{t+1}^{(k)}\right) \ge T_k v(\bar{x}, \bar{x}) - 2 \cdot 2^{-k} - \delta_k. \tag{2.24}$$

Set

$$S_0 = 0, \ S_k = \sum_{i=1}^{k} (T_i + 1) - 1 \text{ for all integers } k \ge 1. \tag{2.25}$$

Define a sequence $\{x_t\}_{t=0}^\infty \subset X$ as follows:

$$x_t = x_t^{(1)}, \ t = 0, \ldots, T_1, \ x_t = x_i^{(k+1)} \tag{2.26}$$

for each integer $k \geq 1$, each $i \in \{0, \ldots, T_{k+1}\}$ and $t = S_k + i + 1$.

It follows from (2.26), (2.18), (2.15), and (2.1) that $\{x_t\}_{t=0}^\infty$ is an (Ω)-program. Relations (2.25), (2.26), (2.18), and (2.15) imply that for each integer $k \geq 1$

$$|v(x_{S_k}, x_{S_k+1}) - v(\bar{x}, \bar{x})| \leq 2 \cdot 2^{-k}. \tag{2.27}$$

By (2.25), (2.26), (2.24), (2.21), and the choice of δ_j, $j = 1, 2, \ldots$ (see (2.15)–(2.18)) for any integer $k \geq 2$

$$\sum_{t=0}^{S_k-1} v(x_t, x_{t+1}) - S_k v(\bar{x}, \bar{x}) = \sum_{j=1}^{k} \left(\sum_{t=0}^{T_j-1} [v\left(x_t^{(j)}, x_{t+1}^{(j)}\right) - v(\bar{x}, \bar{x})] \right)$$

$$+ \sum_{j=1}^{k-1} [v\left(x_{T_j}^{(j)}, x_0^{(j+1)}\right) - v(\bar{x}, \bar{x})] \geq - \sum_{j=1}^{k} (2 \cdot 2^{-j} + \delta_j) - 2 \sum_{j=1}^{k-1} 2^{-j}. \tag{2.28}$$

In view of (2.28) and (2.15), for any integer $k \geq 2$ we have

$$\sum_{t=0}^{S_k-1} v(x_t, x_{t+1}) - S_k v(\bar{x}, \bar{x}) \geq -5 \sum_{j=1}^{k} 2^{-j} \geq -10. \tag{2.29}$$

It follows from (2.29) and Proposition 2.1 that $\{x_t\}_{t=0}^\infty$ is a (v, Ω)-good program. In view of (A3) $\lim_{t\to\infty} \rho(x_t, \bar{x}) = 0$. On the other hand (2.20), (2.25), and (2.26) imply that $\limsup_{t\to\infty} \rho(x_t, \bar{x}) \geq \epsilon$. The contradiction we have reached proves Lemma 2.10.

Lemma 2.11 *Let $\epsilon > 0$ and $M_0 > 0$. Then there exists a natural number T_0 such that for each integer $T \geq T_0$, each (Ω)-program $\{x_t\}_{t=0}^T$ which satisfies*

$$\sum_{t=0}^{T-1} v(x_t, x_{t+1}) \geq T v(\bar{x}, \bar{x}) - M_0 \tag{2.30}$$

and each integer $s \in [0, T - T_0]$ the inequality

$$\min\{\rho(x_i, \bar{x}) : i = s + 1, \ldots, s + T_0\} \leq \epsilon$$

holds.

Proof By Lemma 2.9 there is a natural number T_0 such that the following property holds:

(P0) For each (Ω)-program $\{x_t\}_{t=0}^{T_0}$ which satisfies

$$\sum_{t=0}^{T_0-1} v(x_t, x_{t+1}) \geq T_0 v(\bar{x}, \bar{x}) - M_0 - 2\bar{c}$$

the relation

$$\min\{\rho(x_i, \bar{x}) : i = 1, \ldots, T_0\} \leq \epsilon$$

holds.

Let an integer $T \geq T_0$, let an (Ω)-program $\{x_t\}_{t=0}^{T}$ satisfy (2.30) and let an integer $s \in [0, T - T_0]$. It follows from (2.30) and (A2) that

$$\sum_{t=s}^{s+T_0-1} v(x_t, x_{t+1}) - T_0 v(\bar{x}, \bar{x}) \geq -M_0 - 2\bar{c}.$$

By the inequality above and (P0)

$$\min\{\rho(x_i.\bar{x}) : i = s + 1, \ldots, s + T_0\} \leq \epsilon.$$

Lemma 2.11 is proved.

2.4   Auxiliary Results

We use all the notation, definitions, and assumptions of Sect. 2.1.

Lemma 2.12 *Let ϵ be a positive number and let L be a natural number. Then there exists $\delta > 0$ such that for each*

$$\Omega_t \in \mathcal{E}(\delta), \ t = 0, \ldots, L - 1,$$

each $u_t \in \mathcal{M}, \ t = 0, \ldots, L - 1$ satisfying

$$\|u_t - v\| \leq \delta, \ t = 0, \ldots, L - 1$$

and each $(\{\Omega_t\}_{t=0}^{L-1})$-program $\{x_t\}_{t=0}^{L}$ there exists an (Ω)-program $\{y_t\}_{t=0}^{L}$ such that

$$\rho(x_t, y_t) \leq \epsilon \text{ for all } t = 0, \ldots, L$$

and

$$\sum_{t=0}^{L-1} v(y_t, y_{t+1}) \geq \sum_{t=0}^{L-1} u_t(x_t, x_{t+1}) - \epsilon.$$

Proof Assume that the lemma does not hold. Then for each natural number k there exist sets

$$\Omega_t^{(k)} \in \mathcal{E}(k^{-1}), \ t = 0, \ldots, L - 1, \tag{2.31}$$

functions $u_t^{(k)} \in \mathcal{M}$, $t = 0, \ldots, L - 1$ satisfying

$$\|u_t^{(k)} - v\| \leq k^{-1}, \ t = 0, \ldots, L - 1 \tag{2.32}$$

and an $(\{\Omega_t^{(k)}\}_{t=0}^{L-1})$-program $\{x_t^{(k)}\}_{t=0}^{L}$ such that the following property holds:
(P1) if an (Ω)-program $\{y_t\}_{t=0}^{L}$ satisfies

$$\rho\left(x_t^{(k)}, y_t\right) \leq \epsilon \text{ for all } t = 0, \ldots, L,$$

then

$$\sum_{t=0}^{L-1} v(y_t, y_{t+1}) < \sum_{t=0}^{L-1} u_t\left(x_t^{(k)}, x_{t+1}^{(k)}\right) - \epsilon.$$

Extracting subsequences and reindexing, if necessary, we may assume without loss of generality that for all integers $t = 0, \ldots, L$ there exists

$$x_t = \lim_{k \to \infty} x_t^{(k)}. \tag{2.33}$$

By (2.31) and (2.33), for each integer $t \in \{0, \ldots, L - 1\}$ and each natural number k,

$$\rho_1((x_t, x_{t+1}), \Omega) \leq \rho_1\left((x_t, x_{t+1}), \left(x_t^{(k)}, x_{t+1}^{(k)}\right)\right) + \rho_1\left(\left(x_t^{(k)}, x_{t+1}^{(k)}\right), \Omega\right)$$

$$\leq \rho_1\left((x_t, x_{t+1}), \left(x_t^{(k)}, x_{t+1}^{(k)}\right)\right) + k^{-1} \to 0 \text{ as } k \to \infty$$

and

$$(x_t, x_{t+1}) \in \Omega.$$

Therefore $\{x_t\}_{t=0}^{L}$ is an (Ω)-program.

By (2.33) and upper semicontinuity of the function v, there exists a natural number k_0 such that for each integer $k \geq k_0$,

$$\rho(x_t, x_t^{(k)}) \leq \epsilon, \ t = 0, \ldots, L \tag{2.34}$$

and

$$\sum_{t=0}^{L-1} v(x_t, x_{t+1}) \geq \sum_{t=0}^{L-1} v(x_t^{(k)}, x_{t+1}^{(k)}) - \epsilon/4. \tag{2.35}$$

Choose a natural number $k_1 > k_0$ such that

$$16 k_1^{-1} L < \epsilon. \tag{2.36}$$

Assume that an integer $k \geq k_1$. Then (2.34) and (2.35) hold. In view of (2.32) and (2.35),

$$\sum_{t=0}^{L-1} v(x_t, x_{t+1}) \geq \sum_{t=0}^{L-1} v(x_t^{(k)}, x_{t+1}^{(k)}) - \epsilon/4$$

$$\geq \sum_{t=0}^{L-1} u_t(x_t^{(k)}, x_{t+1}^{(k)}) - \epsilon/4 - L \max\{\|u_t - v\| : t = 0, \ldots, L-1\}$$

$$\geq \sum_{t=0}^{L-1} u_t(x_t^{(k)}, x_{t+1}^{(k)}) - \epsilon/4 - Lk_1^{-1} \geq \sum_{t=0}^{L-1} u_t(x_t^{(k)}, x_{t+1}^{(k)}) - \epsilon/2.$$

When combined with (2.34) this contradicts property (P1). The contradiction we have reached proves Lemma 2.12.

Lemma 2.13 *Let ϵ, M be positive numbers. Then there exists a natural number L such that for each integer $\tilde{L} \geq L$ there exists $\delta > 0$ such that the following assertion holds.*

For each integer $T \in [L, \tilde{L}]$, each

$$\Omega_t \in \mathcal{E}(\delta), \ t = 0, \ldots, T-1, \tag{2.37}$$

each $u_t \in \mathcal{M}$, $t = 0, \ldots, T-1$ satisfying

$$\|u_t - v\| \leq \delta, \ t = 0, \ldots, T-1 \tag{2.38}$$

and each $(\{\Omega_t\}_{t=0}^{T-1})$-program $\{x_t\}_{t=0}^{T}$ satisfying

$$\sum_{t=0}^{T-1} u_t(x_t, x_{t+1}) \geq \sum_{t=0}^{T-1} u_t(\bar{x}, \bar{x}) - M \tag{2.39}$$

the inequality

$$\min\{\rho(x_t, \bar{x}) : \ t = 1, \ldots, T\} \leq \epsilon$$

holds.

Proof We may assume without loss of generality that $\epsilon < 1$. By Lemma 2.11, there exists a natural number L such that the following property holds:

(P2) for each integer $T \geq L$, each (Ω)-program $\{x_t\}_{t=0}^{T}$ which satisfies

$$\sum_{t=0}^{T-1} v(x_t, x_{t+1}) \geq Tv(\bar{x}, \bar{x}) - M - 4$$

and each integer $S \in [0, T-L]$, we have

$$\min\{\rho(x_t, \bar{x}) : \ t = S+1, \ldots, S+L\} \leq \epsilon/4.$$

Let an integer $\tilde{L} \geq L$. By Lemma 2.12, there exists $\delta_0 > 0$ such that the following property holds:

(P3) for each integer $T \in [L, \tilde{L}]$, each

$$\Omega_t \in \mathcal{E}(\delta_0), \ t = 0, \ldots, T - 1,$$

each $u_t \in \mathcal{M}, \ t = 0, \ldots, T - 1$ satisfying

$$\|u_t - v\| \le \delta_0, \ t = 0, \ldots, T - 1$$

and each $(\{\Omega_t\}_{t=0}^{T-1})$-program $\{z_t\}_{t=0}^{T}$ there exists an (Ω)-program $\{y_t\}_{t=0}^{T}$ such that

$$\rho(z_t, y_t) \le \epsilon/4 \text{ for all } t = 0, \ldots, T$$

and

$$\sum_{t=0}^{T-1} v(y_t, y_{t+1}) \ge \sum_{t=0}^{T-1} u_t(z_t, z_{t+1}) - \epsilon/4$$

Set

$$\delta = \min\{\delta_0, \ \epsilon \tilde{L}^{-1}\}. \tag{2.40}$$

Assume that an integer $T \in [L, \tilde{L}]$, (2.37) holds, functions $u_t \in \mathcal{M}, \ t = 0, \ldots, T - 1$ satisfy (2.38) and $(\{\Omega_t\}_{t=0}^{T-1})$-program $\{x_t\}_{t=0}^{T}$ satisfies (2.39).

By (2.37), (2.38), (2.40), and property (P3), there exists an (Ω)-program $\{y_t\}_{t=0}^{T}$ such that

$$\rho(x_t, y_t) \le \epsilon/4 \text{ for all } t = 0, \ldots, T \tag{2.41}$$

and

$$\sum_{t=0}^{T-1} v(y_t, y_{t+1}) \ge \sum_{t=0}^{T-1} u_t(x_t, x_{t+1}) - \epsilon/4. \tag{2.42}$$

In view of (2.38), (2.39), (2.40), and (2.42),

$$\sum_{t=0}^{T-1} v(y_t, y_{t+1}) \ge \sum_{t=0}^{T-1} u_t(x_t, x_{t+1}) - \epsilon/4$$

$$\ge \sum_{t=0}^{T-1} u_t(\bar{x}, \bar{x}) - M - \epsilon/4$$

$$\ge T v(\bar{x}, \bar{x}) - T\delta - M - \epsilon/4$$

$$\ge T v(\bar{x}, \bar{x}) - M - 2\epsilon \ge T v(\bar{x}, \bar{x}) - M - 2. \tag{2.43}$$

It follows from (2.43) and property (P2) that

$$\min\{\rho(y_t, \bar{x}) : \ t = 1, \ldots, L\} \le \epsilon/4.$$

When combined with (2.41) this implies that

$$\min\{\rho(x_t, \bar{x}) : \ t = 1, \ldots, T\} \le \epsilon/2.$$

Lemma 2.13 is proved.

Lemma 2.14 *For each natural number T,*

$$\sigma(v, T, \bar{x}, \bar{x}) = T v(\bar{x}, \bar{x}).$$

Proof Let T be a natural number. Clearly,

$$\sigma(v, T, \bar{x}, \bar{x}) \geq T v(\bar{x}, \bar{x}).$$

Assume that an (Ω)-program $\{y_t\}_{t=0}^{T}$ satisfies

$$y_0 = \bar{x}, \ y_T = \bar{x}. \tag{2.44}$$

For all integers $i > T$ define $y_i \in X$ such that

$$y_{i+T} = y_i \text{ for all integers } i \geq 0. \tag{2.45}$$

It is clear that $\{y_t\}_{t=0}^{\infty}$ is an (Ω)-program. By (2.45) and assumption (A2), for each natural number k,

$$T k v(\bar{x}, \bar{x}) + \bar{c} \geq \sum_{t=0}^{kT-1} v(y_t, y_{t+1}) = k \sum_{t=0}^{T-1} v(y_t, y_{t+1})$$

and

$$k\left(-T v(\bar{x}, \bar{x}) + \sum_{t=0}^{T-1} v(y_t, y_{t+1})\right) \leq \bar{c}$$

for all natural numbers k. This implies that

$$\sum_{t=0}^{T-1} v(y_t, y_{t+1}) \leq T v(\bar{x}, \bar{x})$$

and

$$\sigma(v, T, \bar{x}, \bar{x}) \leq T v(\bar{x}, \bar{x}).$$

Lemma 2.14 is proved.

Lemma 2.15 *Let $\epsilon > 0$. Then there exists $\delta \in (0, \bar{r})$ such that for each natural number T and each $z_0, z_1 \in X$ satisfying*

$$\rho(z_i, \bar{x}) \leq \delta, \ i = 0, 1 \tag{2.46}$$

the inequality

$$|\sigma(v, T, z_0, z_1) - T v(\bar{x}, \bar{x})| \leq \epsilon$$

holds.

Proof In view of (A1), there exists a positive number $\delta < \bar{r}$ such that for each $y, z \in X$ satisfying $\rho(y, \bar{x})$, $\rho(z, \bar{x}) \leq \delta$, we have

$$|v(y, z) - v(\bar{x}, \bar{x})| \leq \epsilon/8. \tag{2.47}$$

Assume that T is a natural number and that $z_0, z_1 \in X$ satisfy (2.46). Define

$$y_0 = z_0, \quad y_i = \bar{x} \text{ for all integers } i \text{ satisfying } 1 \le i < T,$$

$$y_T = z_1. \tag{2.48}$$

By (2.1), (2.46), and (2.48), $\{y_t\}_{t=0}^{\infty}$ is an (Ω)-program and

$$\sigma(v, T, z_0, z_1) \ge \sum_{t=0}^{T-1} v(y_t, y_{t+1}). \tag{2.49}$$

It follows from (2.46), (2.47), and (2.48) that for $i = 0, T-1$,

$$|v(y_i, y_{i+1}) - v(\bar{x}, \bar{x})| \le \epsilon/8.$$

By the inequality above, (2.48) and (2.49),

$$\sigma(v, T, z_0, z_1) \ge T v(\bar{x}, \bar{x}) - \epsilon/4. \tag{2.50}$$

Let an (Ω)-program $\{\tilde{y}_t\}_{t=0}^{\infty}$ satisfy

$$\tilde{y}_0 = z_0, \quad \tilde{y}_T = z_1. \tag{2.51}$$

Set

$$\hat{y}_0 = \bar{x}, \quad \hat{y}_i = \tilde{y}_{i-1}, \ i = 1, \dots, T+1, \ \hat{y}_{T+2} = \bar{x}. \tag{2.52}$$

In view of (2.1), (2.46), (2.51), and (2.52), $\{\hat{y}_t\}_{t=0}^{T+2}$ is an (Ω)-program. It follows from Lemma 2.14, (2.51), and (2.52) that

$$(T+2)v(\bar{x}, \bar{x}) \ge \sum_{t=0}^{T+1} v(\hat{y}_t, \hat{y}_{t+1})$$

$$= \sum_{t=0}^{T-1} v(\tilde{y}_t, \tilde{y}_{t+1}) + v(\bar{x}, z_0) + v(z_1, \bar{x}). \tag{2.53}$$

By (2.46) and (2.47),

$$|v(\bar{x}, z_0) - v(\bar{x}, \bar{x})| \le \epsilon/8,$$

$$|v(z_1, \bar{x}) - v(\bar{x}, \bar{x})| \le \epsilon/8. \tag{2.54}$$

It follows from (2.53) and (2.54) that

$$(T+2)v(\bar{x}, \bar{x}) \ge \sum_{t=0}^{T-1} v(\tilde{y}_t, \tilde{y}_{t+1}) + 2v(\bar{x}, \bar{x}) - \epsilon/4$$

and

$$\sum_{t=0}^{T-1} v(\tilde{y}_t, \tilde{y}_{t+1}) \le T v(\bar{x}, \bar{x}) + \epsilon/4.$$

Since this inequality holds for any (Ω)-program $\{\tilde{y}_t\}_{t=0}^{T}$ satisfying (2.51) we have

$$\sigma(v, T, z_0, z_1) \le T v(\bar{x}, \bar{x}) + \epsilon/4.$$

When combined with (2.50) this inequality completes the proof of Lemma 2.15.

Lemma 2.16 *Let ϵ be a positive number. Then there exists $\delta \in (0, \bar{\lambda})$ such that for each natural number L there exists $\delta_0 \in (0, \delta)$ such that the following assertion holds.*

For each natural number $T \leq L$, each

$$\Omega_t \in \mathcal{E}(\delta_0), \; t = 0, \ldots, T - 1, \tag{2.55}$$

each $u_t \in \mathcal{M}$, $t = 0, \ldots, T - 1$ satisfying

$$\|u_t - v\| \leq \delta_0, \; t = 0, \ldots, T - 1 \tag{2.56}$$

and each $(\{\Omega_t\}_{t=0}^{T-1})$-program $\{x_t\}_{t=0}^{T}$ satisfying

$$\rho(x_0, \bar{x}), \; \rho(x_T, \bar{x}) \leq \delta \tag{2.57}$$

and

$$\sum_{t=0}^{T-1} u_t(x_t, x_{t+1}) \geq \sigma(\{u_t\}_{t=0}^{T-1}, \{\Omega_t\}_{t=0}^{T-1}, 0, T, x_0, x_T) - \delta \tag{2.58}$$

the inequality $\rho(x_t, \bar{x}) \leq \epsilon$ holds for all $t = 0, \ldots, T$.

Proof By Lemma 2.10, there exists a positive number

$$\gamma < \min\{\epsilon/4, \; \bar{\lambda}/4\}$$

such that the following property holds:
   (P4) for each integer $T \geq 1$ and each (Ω)-program $\{x_t\}_{t=0}^{T}$ which satisfies

$$\rho(x_0, \bar{x}), \; \rho(x_T, \bar{x}) \leq 2\gamma,$$

$$\sum_{t=0}^{T-1} v(x_t, x_{t+1}) \geq \sigma(v, T, x_0, x_T) - 2\gamma$$

the inequality $\rho(x_t, \bar{x}) \leq \epsilon/4$ holds for all $t = 0, \ldots, T$.
   By Lemma 2.15, there exists $\delta \in (0, \gamma/4)$ such that the following property holds:
   (P5) for each natural number T and each $z_0, z_1 \in X$ satisfying

$$\rho(z_i, \bar{x}) \leq 2\delta, \; i = 0, 1,$$

we have

$$|\sigma(v, T, z_0, z_1) - Tv(\bar{x}, \bar{x})| \leq \gamma/4.$$

   Let L be a natural number. By Lemma 2.12, there exists $\delta_1 \in (0, \delta)$ such that the following property holds:

(P6) for each integer $T \in [1, L]$, each

$$\Omega_t \in \mathcal{E}(\delta_1), \ t = 0, \dots, T - 1,$$

each $u_t \in \mathcal{M}$, $t = 0, \dots, T - 1$ satisfying

$$\|u_t - v\| \leq \delta_1, \ t = 0, \dots, T - 1$$

and each $(\{\Omega_t\}_{t=0}^{T-1})$-program $\{x_t\}_{t=0}^{T}$ there exists an (Ω)-program $\{y_t\}_{t=0}^{T}$ such that

$$\rho(x_t, y_t) \leq \delta/4 \text{ for all } t = 0, \dots, T$$

and

$$\sum_{t=0}^{T-1} v(y_t, y_{t+1}) \geq \sum_{t=0}^{T-1} u_t(x_t, x_{t+1}) - \delta/4.$$

Set

$$\delta_0 = (2\,L)^{-1}\delta_1. \tag{2.59}$$

Assume that an integer $T \in [1, L]$, (2.55) holds, functions $u_t \in \mathcal{M}$, $t = 0, \dots, T - 1$ satisfy (2.56) and that an $(\{\Omega_t\}_{t=0}^{T-1})$-program $\{x_t\}_{t=0}^{T}$ satisfies (2.57) and (2.58).

By (2.55), (2.56), (2.59), and property (P6), there exists an (Ω)-program $\{y_t\}_{t=0}^{T}$ such that

$$\rho(x_t, y_t) \leq \delta/4 \text{ for all } t = 0, \dots, T \tag{2.60}$$

and

$$\sum_{t=0}^{T-1} v(y_t, y_{t+1}) \geq \sum_{t=0}^{T-1} u_t(x_t, x_{t+1}) - \delta/4. \tag{2.61}$$

In view of (2.58) and (2.61),

$$\sum_{t=0}^{T-1} v(y_t, y_{t+1}) \geq \sigma\left(\{u_t\}_{t=0}^{T-1}, \{\Omega_t\}_{t=0}^{T-1}, 0, T, x_0, x_T\right) - \delta - \delta/4. \tag{2.62}$$

Set

$$\tilde{x}_0 = x_0, \ \tilde{x}_t = \bar{x} \text{ for all integers } t \text{ satisfying } 1 \leq i < T,$$

$$\tilde{x}_T = x_T. \tag{2.63}$$

In view of (2.4), (2.55), (2.57), and (2.63), $\{\tilde{x}_t\}_{t=0}^{T}$ is an $(\{\Omega_t\}_{t=0}^{T-1})$-program. It follows from (2.57), (2.63), and property (P5) that

$$|v(\tilde{x}_i, \tilde{x}_{i+1}) - v(\bar{x}, \bar{x})| \leq \gamma/4, \ i = 0, \ T - 1. \tag{2.64}$$

By (2.56), (2.59), (2.62), and (2.63),

$$\sum_{t=0}^{T-1} v(y_t, y_{t+1}) \geq \sum_{t=0}^{T-1} u_t(\tilde{x}_t, \tilde{x}_{t+1}) - \delta - \delta/4$$

$$\geq -\delta - \delta/4 + \sum_{t=0}^{T-1} v(\tilde{x}_t, \tilde{x}_{t+1})$$

$$- T \max\{\|u_t - v\| : t = 0, \ldots, T-1\}$$

$$\geq \sum_{t=0}^{T-1} v(\tilde{x}_t, \tilde{x}_{t+1}) - \delta - \delta/4 - L\delta_0$$

$$\geq \sum_{t=0}^{T-1} v(\tilde{x}_t, \tilde{x}_{t+1}) - \delta - \delta/4 - \delta/2. \tag{2.65}$$

It follows from (2.63), (2.64), and (2.65) that

$$\sum_{t=0}^{T-1} v(y_t, y_{t+1}) \geq T v(\bar{x}, \bar{x}) - \gamma/2 - 2\delta. \tag{2.66}$$

In view of (2.60), for $i = 0, T$,

$$\rho(y_i, \bar{x}) \leq \rho(y_i, x_i) + \rho(x_i, \bar{x}) \leq \delta/4 + \delta. \tag{2.67}$$

By (2.67) and property (P5),

$$|\sigma(v, T, y_0, y_T) - T v(\bar{x}, \bar{x})| \leq \gamma/4. \tag{2.68}$$

It follows from (2.66) and (2.68) that

$$\sum_{t=0}^{T-1} v(y_t, y_{t+1}) \geq \sigma(v, T, y_0, y_T) - \gamma/2 - \gamma/4 - 2\delta$$

$$\geq \sigma(v, T, y_0, y_T) - 2\gamma. \tag{2.69}$$

In view of (2.67), (2.69), and property (P4),

$$\rho(y_t, \bar{x}) \leq \epsilon/4 \text{ for all } t = 0, \ldots, T. \tag{2.70}$$

By (2.60) and (2.70), for all $t = 0, \ldots, T$,

$$\rho(x_t, \bar{x}) \leq \rho(x_t, y_t) + \rho(y_t, \bar{x}) \leq \delta/4 + \epsilon/4 < \epsilon.$$

This completes the proof of Lemma 2.16.

In order to prove our stability results we need the following two lemmas.

Lemma 2.17 *Let ϵ, M_0 be positive numbers and l_1, l_2 be natural numbers. Then there exist $\delta > 0$ and a natural number $L > l_1 + l_2$ such that for each integer $T \geq L$, each*

$$\Omega_t \in \mathcal{E}(\delta), \ t = 0, \ldots, T - 1,$$

each $u_t \in \mathcal{M}$, $t = 0, \ldots, T - 1$ satisfying

$$\|u_t - v\| \leq \delta, \ t = 0, \ldots, T - 1,$$

each $(\{\Omega_t\}_{t=0}^{T-1})$-program $\{x_t\}_{t=0}^{T}$ satisfying

$$x_T \in Y(\{\Omega_t\}_{t=T-l_2}^{T-1}, T - l_2, T)$$

and

$$\sigma(\{u_t\}_{t=0}^{T-1}, \{\Omega_t\}_{t=0}^{T-1}, 0, T, x_0, x_T) \leq \sum_{t=0}^{T-1} u_t(x_t, x_{t+1}) + M_0$$

and each integer $S \in [0, T - L]$ satisfying

$$x_S \in \bar{Y}(\{\Omega_t\}_{t=S}^{S+l_1-1}, S, S + l_1)$$

the inequality

$$\min\{\rho(x_t, \bar{x}) : \ t = S + 1, \ldots, S + L\} \leq \epsilon$$

holds.

Lemma 2.18 *Let ϵ, M_0 be positive numbers and l_1 be a natural number. Then there exists $\delta > 0$ and a natural number $L > l_1$ such that for each integer $T \geq L$, each*

$$\Omega_t \in \mathcal{E}(\delta), \ t = 0, \ldots, T - 1,$$

each $u_t \in \mathcal{M}$, $t = 0, \ldots, T - 1$ satisfying

$$\|u_t - v\| \leq \delta, \ t = 0, \ldots, T - 1,$$

each $(\{\Omega_t\}_{t=0}^{T-1})$-program $\{x_t\}_{t=0}^{T}$ satisfying

$$\sigma(\{u_t\}_{t=0}^{T-1}, \{\Omega_t\}_{t=0}^{T-1}, 0, T, x_0) \leq \sum_{t=0}^{T-1} u_t(x_t, x_{t+1}) + M_0$$

and each integer $S \in [0, T - L]$ satisfying

$$x_S \in \bar{Y}(\{\Omega_t\}_{t=S}^{S+l_1-1}, S, S + l_1)$$

the inequality

$$\min\{\rho(x_t, \bar{x}) : \ t = S + 1, \ldots, S + L\} \leq \epsilon$$

holds.

Proof We prove Lemmas 2.17 and 2.18 simultaneously. We may assume without loss of generality that

$$\epsilon < \bar{\lambda}. \tag{2.71}$$

By Lemma 2.13, there exists a natural number L_0 and $\delta \in (0, \epsilon)$ such that the following property holds:
   (P7) for each

$$\Omega_t \in \mathcal{E}(\delta), \ t = 0, \ldots, L_0 - 1,$$

each $u_t \in \mathcal{M}, \ t = 0, \ldots, L_0 - 1$ satisfying

$$\|u_t - v\| \le \delta, \ t = 0, \ldots, L_0 - 1$$

and each $(\{\Omega_t\}_{t=0}^{L_0-1})$-program $\{x_t\}_{t=0}^{L_0}$ satisfying

$$\sum_{t=0}^{L_0-1} u_t(x_t, x_{t+1}) \ge \sum_{t=0}^{L_0-1} u_t(\bar{x}, \bar{x}) - 1$$

the inequality

$$\min\{\rho(x_t, \bar{x}) : \ t = 1, \ldots, L_0\} \le \epsilon$$

holds.
   In the case of Lemma 2.18 set $l_2 = 1$.
   Choose a natural number k_0 such that

$$k_0 > 2(\|v\| + 1)(L_0 + l_1 + l_2 + 1) + M_0 + 1. \tag{2.72}$$

Set

$$L = L_0 k_0. \tag{2.73}$$

Let an integer $T \ge L$,

$$\Omega_t \in \mathcal{E}(\delta), \ t = 0, \ldots, T - 1, \tag{2.74}$$

functions $u_t \in \mathcal{M}, \ t = 0, \ldots, T - 1$ satisfy

$$\|u_t - v\| \le \delta, \ t = 0, \ldots, T - 1, \tag{2.75}$$

$(\{\Omega_t\}_{t=0}^{T-1})$-program $\{x_t\}_{t=0}^{T}$ satisfies

$$x_T \in Y(\{\Omega_t\}_{t=T-l_2}^{T-1}, T - l_2, T) \tag{2.76}$$

and

$$\sigma(\{u_t\}_{t=0}^{T-1}, \{\Omega_t\}_{t=0}^{T-1}, 0, T, x_0, x_T) \le \sum_{t=0}^{T-1} u_t(x_t, x_{t+1}) + M_0 \tag{2.77}$$

in the case of Lemma 2.17 and satisfies

$$\sigma(\{u_t\}_{t=0}^{T-1}, \{\Omega_t\}_{t=0}^{T-1}, 0, T, x_0) \leq \sum_{t=0}^{T-1} u_t(x_t, x_{t+1}) + M_0 \qquad (2.78)$$

in the case of Lemma 2.18. Assume that an integer

$$S \in [0, T - L] \qquad (2.79)$$

satisfies

$$x_S \in \bar{Y}\left(\{\Omega_t\}_{t=S}^{S+l_1-1}, S, S+l_1\right). \qquad (2.80)$$

In order to complete the proof of Lemmas 2.17 and 2.18 it is sufficient to show that

$$\min\{\rho(x_t, \bar{x}) : t = S + 1, \ldots, S + L\} \leq \epsilon.$$

Assume the contrary. Then

$$\rho(x_t, \bar{x}) > \epsilon \text{ for all } t = S + 1, \ldots, S + L. \qquad (2.81)$$

There are two cases:
(1) There is an integer $S_0 \in (S, T]$ such that

$$\rho(x_{S_0}, \bar{x}) \leq \epsilon. \qquad (2.82)$$

(2)

$$\rho(x_t, \bar{x}) > \epsilon \text{ for all } t = S + 1, \ldots, T. \qquad (2.83)$$

Assume that the case (1) holds. In view of (2.81) and (2.82),

$$S_0 > S + L. \qquad (2.84)$$

We may assume without loss of generality that

$$\rho(x_t, \bar{x}) > \epsilon \text{ for all } t = S + 1, \ldots, S_0 - 1. \qquad (2.85)$$

In view of (2.80), there exists an $\left(\{\Omega_t\}_{t=S}^{S+l_1-1}\right)$-program $\{y_t\}_{t=S}^{S+l_1}$ such that

$$y_S = x_S, \quad y_{S+l_1} = \bar{x}. \qquad (2.86)$$

Set

$$y_{S_0} = x_{S_0}, \quad y_t = \bar{x}, \quad t = S + l_1 + 1, \ldots, S_0 - 1. \qquad (2.87)$$

By (2.4), (2.71), (2.82), (2.86), and (2.87),

$$\{y_t\}_{t=S}^{S_0} \text{ is an } \left(\{\Omega_t\}_{t=S}^{S_0-1}\right)\text{—program.} \qquad (2.88)$$

It follows from (2.75), (2.77), (2.78), (2.86), (2.87), and (2.88) that

$$
\sum_{t=S}^{S_0-1} u_t(x_t, x_{t+1}) \geq \sum_{t=S}^{S_0-1} u_t(y_t, y_{t+1}) - M_0
$$

$$
\geq \sum_{t=S}^{S_0-1} u_t(\bar{x}, \bar{x}) - 2 \sum_{t=S}^{S+l_1-1} \|u_t\| - 2\|u_{S_0-1}\| - M_0
$$

$$
\geq \sum_{t=S}^{S_0-1} u_t(\bar{x}, \bar{x}) - M_0 - 2(l_1 + 1)(\|v\| + 1). \tag{2.89}
$$

There exists a natural number k such that

$$
S_0 - S \in (kL_0, (k+1)L_0]. \tag{2.90}
$$

By (2.73), (2.84), and (2.90),

$$
k \geq k_0. \tag{2.91}
$$

It follows from (2.74), (2.75), (2.85), (2.90), (2.91), and property (P7) that for each integer $j \in [0, k-1]$,

$$
\sum_{t=S+jL_0}^{S+(j+1)L_0-1} u_t(x_t, x_{t+1}) < \sum_{t=S+jL_0}^{S+(j+1)L_0-1} u_t(\bar{x}, \bar{x}) - 1. \tag{2.92}
$$

By (2.71), (2.75), (2.89), (2.90), (2.91), and (2.92),

$$
-M_0 - 2(l_1 + 1)(\|v\| + 1) \leq \sum_{t=S}^{S_0-1} u_t(x_t, x_{t+1}) - \sum_{t=S}^{S_0-1} u_t(\bar{x}, \bar{x})
$$

$$
\leq \sum_{j=0}^{k-1} \left[\sum_{t=S+jL_0}^{S+(j+1)L_0-1} (u_t(x_t, x_{t+1}) - u_t(\bar{x}, \bar{x})) \right]
$$

$$
+ 2L_0 \max\{\|u_t\| : t = 0, \ldots, S_0 - 1\}
$$

$$
\leq -k + 2L_0(\|v\| + 1) \leq -k_0 + 2L_0(\|v\| + 1)
$$

and

$$
k_0 \leq 2(\|v\| + 1)(L_0 + l_1 + l_2 + 1) + M_0.
$$

This contradicts (2.72). The contradiction we have reached proves that case (1) does not hold. Therefore case (2) holds and (2.83) is true.

In view of (2.80), there exists an $(\{\Omega_t\}_{t=S}^{S+l_1-1})$-program $\{y_t\}_{t=S}^{S+l_1}$ such that (2.86) holds. In the case of Lemma 2.18 set

$$
y_t = \bar{x} \quad \text{for all integers } t \in [S + l_1 + 1, T] \tag{2.93}
$$

and then in view of (2.86) and (2.93), $\{y_t\}_{t=S}^T$ is an $(\{\Omega_t\}_{t=S}^{T-1})$-program.

In the case of Lemma 2.17 it follows from (2.76) that there exist $y_t \in X$, $t = T - l_2, \ldots, T$ such that

$$\{y_t\}_{t=T-l_2}^T \text{ is an } (\{\Omega_t\}_{t=T-l_2}^{T-1})\text{-program} \tag{2.94}$$

which satisfies

$$y_{T-l_2} = \bar{x}, \quad y_T = x_T. \tag{2.95}$$

In the case of Lemma 2.17 we set

$$y_t = \bar{x} \text{ for all integers } t \text{ satisfying } S + l_1 < t < T - l_2 \tag{2.96}$$

and then in view of (2.86) and (2.94)–(2.96), $\{y_t\}_{t=S}^T$ is also an $(\{\Omega_t\}_{t=S}^{T-1})$-program.

It follows from (2.77), (2.78), (2.86), and (2.95) that in both cases

$$\sum_{t=S}^{T-1} u_t(x_t, x_{t+1}) \geq \sum_{t=S}^{T-1} u_t(y_t, y_{t+1}) - M_0. \tag{2.97}$$

By (2.86), (2.93), and (2.95),

$$y_t = \bar{x} \text{ for all integers } t = S + l_1, \ldots, T - l_2. \tag{2.98}$$

By (2.75), (2.97), and (2.98),

$$\sum_{t=S}^{T-1} u_t(x_t, x_{t+1}) \geq \sum_{t=S}^{T-1} u_t(y_t, y_{t+1}) - M_0$$

$$\geq \sum_{t=S}^{T-1} u_t(\bar{x}, \bar{x}) - 2(l_1 + l_2)\max\{\|u_t\| : t = 0, \ldots, T\} - M_0$$

$$\geq \sum_{t=S}^{T-1} u_t(\bar{x}, \bar{x}) - M_0 - 2(l_1 + l_2)(\|v\| + 1). \tag{2.99}$$

There exists a natural number k such that

$$T - S \in [kL_0, (k+1)L_0). \tag{2.100}$$

By (2.79), (2.93), and (2.100),

$$k \geq k_0.$$

It follows from the inequality above, (2.74), (2.75), (2.83), (2.100), and property (P7) that for each integer $j \in [0, k-1]$,

$$\sum_{t=S+jL_0}^{S+(j+1)L_0-1} u_t(x_t, x_{t+1}) < \sum_{t=S+jL_0}^{S+(j+1)L_0-1} u_t(\bar{x}, \bar{x}) - 1.$$

By the inequality above, (2.75), (2.99), (2.100), and the inequality $k \geq k_0$,

$$-M_0 - 2(l_1 + l_2)(\|v\| + 1) \leq \sum_{t=S}^{T-1} u_t(x_t, x_{t+1}) - \sum_{t=S}^{T-1} u_t(\bar{x}, \bar{x})$$

$$\leq \sum_{j=0}^{k-1} [\sum_{t=S+jL_0}^{S+(j+1)L_0-1} (u_t(x_t, x_{t+1}) - u_t(\bar{x}, \bar{x}))]$$

$$+ 2L_0 \max\{\|u_t\| : t = 0, \ldots, T - 1\}$$

$$\leq -k + 2L_0(\|v\| + 1) \leq -k_0 + 2L_0(\|v\| + 1)$$

and

$$k_0 \leq 2(\|v\| + 1)(L_0 + l_1 + l_2 + 1) + M_0.$$

This contradicts (2.72). The contradiction we have reached completes the proof of Lemmas 2.17 and 2.18.

2.5 Proofs of Theorems 2.2 and 2.3

We prove Theorems 2.2 and 2.3 simultaneously.

By Lemma 2.16, there exists $\gamma \in (0, \bar{\lambda})$ such that the following property holds:

(P8) for each natural number L there exists $\gamma_L \in (0, \gamma)$ such that for each natural number $T \leq L$, each

$$\Omega_t \in \mathcal{E}(\gamma_L), \ t = 0, \ldots, T - 1,$$

each $u_t \in \mathcal{M}, \ t = 0, \ldots, T - 1$ satisfying

$$\|u_t - v\| \leq \gamma_L, \ t = 0, \ldots, T - 1$$

and each $(\{\Omega_t\}_{t=0}^{T-1})$-program $\{y_t\}_{t=0}^{T}$ satisfying

$$\rho(y_0, \bar{x}), \ \rho(y_T, \bar{x}) \leq \gamma$$

and

$$\sum_{t=0}^{T-1} u_t(y_t, y_{t+1}) \geq \sigma(\{u_t\}_{t=0}^{T-1}, \{\Omega_t\}_{t=0}^{T-1}, 0, T, y_0, y_T) - \gamma$$

the inequality $\rho(y_t, \bar{x}) \leq \epsilon$ holds for all $t = 0, \ldots, T$.

In the case of Theorem 2.3 set $l_2 = 1$.

By Lemmas 2.17 and 2.18 (with $\epsilon = \gamma$ and $M_0 = 1$), there exist

$$\tilde{\gamma} \in (0, \gamma)$$

and a natural number

$$L > l_1 + l_2$$

such that the following properties hold:

(i) For each integer $T \geq L$, each

$$\Omega_t \in \mathcal{E}(\tilde{\gamma}), \ t = 0, \ldots, T-1, \tag{2.101}$$

each $u_t \in \mathcal{M}, \ t = 0, \ldots, T-1$ satisfying

$$\|u_t - v\| \leq \tilde{\gamma}, \ t = 0, \ldots, T-1, \tag{2.102}$$

each $(\{\Omega_t\}_{t=0}^{T-1})$-program $\{z_t\}_{t=0}^{T}$ satisfying

$$z_T \in Y(\{\Omega_t\}_{t=T-l_2}^{T-1}, T-l_2, T)$$

and

$$\sigma(\{u_t\}_{t=0}^{T-1}, \{\Omega_t\}_{t=0}^{T-1}, 0, T, z_0, z_T) \leq \sum_{t=0}^{T-1} u_t(z_t, z_{t+1}) + 1$$

and each integer $S \in [0, T-L]$ satisfying

$$z_S \in \bar{Y}(\{\Omega_t\}_{t=S}^{S+l_1-1}, S, S+l_1) \tag{2.103}$$

the inequality

$$\min\{\rho(z_t, \bar{x}) : \ t = S+1, \ldots, S+L\} \leq \gamma \tag{2.104}$$

holds.

(ii) For each integer $T \geq L$, each $\Omega_t \subset X \times X, t = 0, \ldots, T-1$, satisfying (2.101), each $u_t \in \mathcal{M}, \ t = 0, \ldots, T-1$ satisfying (2.102), each $(\{\Omega_t\}_{t=0}^{T-1})$-program $\{z_t\}_{t=0}^{T}$ satisfying

$$\sigma(\{u_t\}_{t=0}^{T-1}, \{\Omega_t\}_{t=0}^{T-1}, 0, T, z_0) \leq \sum_{t=0}^{T-1} u_t(z_t, z_{t+1}) + 1$$

and each integer $S \in [0, T-L]$ satisfying (2.103) inequality (2.104) holds.

By (P8) there exists $\delta \in (0, \tilde{\gamma})$ such that the following property holds:

(P9) for each natural number $T \leq 2L+4$, each

$$\Omega_t \in \mathcal{E}(\delta), \ t = 0, \ldots, T-1,$$

each $u_t \in \mathcal{M}, \ t = 0, \ldots, T-1$ satisfying

$$\|u_t - v\| \leq \delta, \ t = 0, \ldots, T-1$$

and each $(\{\Omega_t\}_{t=0}^{T-1})$-program $\{y_t\}_{t=0}^{T}$ satisfying

$$\rho(y_0, \bar{x}), \ \rho(y_T, \bar{x}) \leq \gamma$$

and

$$\sum_{t=0}^{T-1} u_t(y_t, y_{t+1}) \geq \sigma(\{u_t\}_{t=0}^{T-1}, \{\Omega_t\}_{t=0}^{T-1}, 0, T, y_0, y_T) - \gamma$$

the inequality $\rho(y_t, \bar{x}) \leq \epsilon$ holds for all $t = 0, \ldots, T$.

Assume that an integer $T > 2L$,

$$\Omega_t \in \mathcal{E}(\delta), \ t = 0, \ldots, T - 1, \tag{2.105}$$

$u_t \in \mathcal{M}, \ t = 0, \ldots, T - 1$ satisfy

$$\|u_t - v\| \le \delta, \ t = 0, \ldots, T - 1, \tag{2.106}$$

and that an $(\{\Omega_t\}_{t=0}^{T-1})$-program $\{x_t\}_{t=0}^{T}$ satisfies

$$x_0 \in \bar{Y}(\{\Omega_t\}_{t=0}^{l_1-1}, 0, l_1). \tag{2.107}$$

Moreover, assume that in the case of Theorem 2.2

$$x_T \in Y(\{\Omega_t\}_{t=T-l_2}^{T-1}, T - l_2, T), \tag{2.108}$$

$$\sigma(\{u_t\}_{t=0}^{T-1}, \{\Omega_t\}_{t=0}^{T-1}, 0, T, x_0, x_T) \le \sum_{t=0}^{T-1} u_t(x_t, x_{t+1}) + \delta \tag{2.109}$$

and that in the case of Theorem 2.3,

$$\sigma(\{u_t\}_{t=0}^{T-1}, \{\Omega_t\}_{t=0}^{T-1}, 0, T, x_0) \le \sum_{t=0}^{T-1} u_t(x_t, x_{t+1}) + \delta. \tag{2.110}$$

By (2.1), (2.4), (2.105)–(2.109), and the inequality $\gamma < \bar{\lambda}$, applying by induction property (i) in the case of Theorem 2.2 and applying by induction property (ii) in the case of Theorem 2.3, we obtain a finite sequence of integer $S_i, \ i = 0, \ldots, q$ such that

$$0 \le S_0 \le L, \ T - L < S_q \le T,$$

$$1 \le S_{i+1} - S_i \le L, \ i = 0, \ldots, q - 1,$$

$$\rho(x_{S_i}, \bar{x}) \le \gamma, \ i = 0, \ldots, q.$$

If $\rho(x_0, \bar{x}) \le \delta$, we may assume that $S_0 = 0$ and if $\rho(x_T, \bar{x}) \le \delta$, we may assume that $S_q = T$. Set

$$\tau_1 = S_0, \ \tau_2 = S_q.$$

Let an integer $t \in [\tau_1, \tau_2]$. Then there exists an integer $i \in [0, q - 1]$ such that

$$t \in [S_i, S_{i+1}].$$

By the inclusion above, the choice of $S_i, \ i = 0, \ldots, q$, (2.105), (2.106), (2.109), (2.110), and property (P9),

$$\rho(x_t, \bar{x}) \le \epsilon.$$

This completes the proof of Theorems 2.2 and 2.3.

2.6   Proofs of Theorems 2.4 and 2.5

We prove Theorems 2.4 and 2.5 simultaneously.

By Lemma 2.16, there exists $\gamma \in (0, \bar{\lambda})$ such that the following property holds:

(P10) For each natural number L there exists $\gamma_L \in (0, \gamma)$ such that for each natural number $T \leq L$, each

$$\Omega_t \in \mathcal{E}(\gamma_L), \ t = 0, \ldots, T - 1,$$

each $u_t \in \mathcal{M}, \ t = 0, \ldots, T - 1$ satisfying

$$\|u_t - v\| \leq \gamma_L, \ t = 0, \ldots, T - 1$$

and each $(\{\Omega_t\}_{t=0}^{T-1})$-program $\{y_t\}_{t=0}^{T}$ satisfying

$$\rho(y_0, \bar{x}), \ \rho(y_T, \bar{x}) \leq \gamma$$

and

$$\sum_{t=0}^{T-1} u_t(y_t, y_{t+1}) \geq \sigma(\{u_t\}_{t=0}^{T-1}, \{\Omega_t\}_{t=0}^{T-1}, 0, T, y_0, y_T) - \gamma$$

the inequality $\rho(y_t, \bar{x}) \leq \epsilon$ holds for all $t = 0, \ldots, T$.

In the case of Theorem 2.5 set $l_2 = 1$.

By Lemmas 2.17 and 2.18 (with $\epsilon = \gamma$ and $M_0 = M + 1$), there exist

$$\tilde{\gamma} \in (0, \gamma)$$

and a natural number

$$L_0 > l_1 + l_2$$

such that the following properties hold:

(i) For each integer $T \geq L_0$, each

$$\Omega_t \in \mathcal{E}(\tilde{\gamma}), \ t = 0, \ldots, T - 1, \tag{2.111}$$

each $u_t \in \mathcal{M}, \ t = 0, \ldots, T - 1$ satisfying

$$\|u_t - v\| \leq \tilde{\gamma}, \ t = 0, \ldots, T - 1, \tag{2.112}$$

each $(\{\Omega_t\}_{t=0}^{T-1})$-program $\{z_t\}_{t=0}^{T}$ satisfying

$$z_T \in Y(\{\Omega_t\}_{t=T-l_2}^{T-1}, T - l_2, T)$$

and

$$\sigma(\{u_t\}_{t=0}^{T-1}, \{\Omega_t\}_{t=0}^{T-1}, 0, T, z_0, z_T) \leq \sum_{t=0}^{T-1} u_t(z_t, z_{t+1}) + 1 + M$$

and each integer $S \in [0, T - L_0]$ satisfying

$$z_S \in \bar{Y}(\{\Omega_t\}_{t=S}^{S+l_1-1}, S, S + l_1) \tag{2.113}$$

the inequality

$$\min\{\rho(z_t, \bar{x}) : \ t = S+1, \ldots, S + L_0\} \leq \gamma \tag{2.114}$$

holds.

(ii) For each integer $T \geq L_0$, each $\Omega_t \subset X \times X$, $t = 0, \ldots, T-1$, satisfying (2.111), each $u_t \in \mathcal{M}$, $t = 0, \ldots, T-1$ satisfying (2.112), each $(\{\Omega_t\}_{t=0}^{T-1})$-program $\{z_t\}_{t=0}^{T}$ satisfying

$$\sigma(\{u_t\}_{t=0}^{T-1}, \{\Omega_t\}_{t=0}^{T-1}, 0, T, z_0) \leq \sum_{t=0}^{T-1} u_t(z_t, z_{t+1}) + 1 + M$$

and each integer $S \in [0, T - L_0]$ satisfying (2.113) inequality (2.114) holds.

By (P10) there exists $\delta \in (0, \tilde{\gamma})$ such that the following property holds:

(P11) For each natural number $T \leq 2L_0 + 4$, each

$$\Omega_t \in \mathcal{E}(\delta), \ t = 0, \ldots, T-1,$$

each $u_t \in \mathcal{M}$, $t = 0, \ldots, T-1$ satisfying

$$\|u_t - v\| \leq \delta, \ t = 0, \ldots, T-1$$

and each $(\{\Omega_t\}_{t=0}^{T-1})$-program $\{y_t\}_{t=0}^{T}$ satisfying

$$\rho(y_0, \bar{x}), \ \rho(y_T, \bar{x}) \leq \gamma$$

and

$$\sum_{t=0}^{T-1} u_t(y_t, y_{t+1}) \geq \sigma(\{u_t\}_{t=0}^{T-1}, \{\Omega_t\}_{t=0}^{T-1}, 0, T, y_0, y_T) - \gamma$$

the inequality $\rho(y_t, \bar{x}) \leq \epsilon$ holds for all $t = 0, \ldots, T$.

Choose a natural number

$$L > (4 + \gamma^{-1}(1 + M))(L_0 + 1). \tag{2.115}$$

Assume that an integer $T > L$,

$$\Omega_t \in \mathcal{E}(\delta), \ t = 0, \ldots, T-1, \tag{2.116}$$

$u_t \in \mathcal{M}$, $t = 0, \ldots, T-1$ satisfy

$$\|u_t - v\| \leq \delta, \ t = 0, \ldots, T-1, \tag{2.117}$$

and that an $(\{\Omega_t\}_{t=0}^{T-1})$-program $\{x_t\}_{t=0}^{T}$ satisfies

$$x_0 \in \bar{Y}(\{\Omega_t\}_{t=0}^{l_1-1}, 0, l_1). \tag{2.118}$$

Moreover, assume that in the case of Theorem 2.4

$$x_T \in Y(\{\Omega_t\}_{t=T-l_2}^{T-1}, T - l_2, T), \tag{2.119}$$

$$\sigma(\{u_t\}_{t=0}^{T-1}, \{\Omega_t\}_{t=0}^{T-1}, 0, T, x_0, x_T) \le \sum_{t=0}^{T-1} u_t(x_t, x_{t+1}) + M \tag{2.120}$$

and that in the case of Theorem 2.5

$$\sigma(\{u_t\}_{t=0}^{T-1}, \{\Omega_t\}_{t=0}^{T-1}, 0, T, x_0) \le \sum_{t=0}^{T-1} u_t(x_t, x_{t+1}) + M. \tag{2.121}$$

By (2.115), (2.116)–(2.121), and the inequality $\delta < \tilde{\gamma}$, applying by induction property (i) in the case of Theorem 2.4 and applying by induction propert (ii) in the case of Theorem 2.5, we obtain a finite sequence of integers $S_i, i = 0, \dots, q$ such that

$$0 \le S_0 \le L_0, \ T - L_0 < S_q \le T, \tag{2.122}$$

$$1 \le S_{i+1} - S_i \le L_0, \ i = 0, \dots, q - 1, \tag{2.123}$$

$$\rho(x_{S_i}, \bar{x}) \le \gamma, \ i = 0, \dots, q. \tag{2.124}$$

By (2.120), (2.121), and (2.122),

$$M \ge \sigma(\{u_t\}_{t=0}^{T-1}, \{\Omega_t\}_{t=0}^{T-1}, 0, T, x_0, x_T) - \sum_{t=0}^{T-1} u_t(x_t, x_{t+1})$$

$$\ge \sum_{i=0}^{q-1} \left[\sigma(\{u_t\}_{t=S_i}^{S_{i+1}-1}, \{\Omega_t\}_{t=S_i}^{S_{i+1}-1}, S_i, S_{i+1}, x_{S_i}, x_{S_{i+1}}) - \sum_{t=S_i}^{S_{i+1}-1} u_t(x_t, x_{t+1}) \right]. \tag{2.125}$$

Set

$$E = \{i \in \{0, \dots, q-1\} :$$

$$\sigma(\{u_t\}_{t=S_i}^{S_{i+1}-1}, \{\Omega_t\}_{t=S_i}^{S_{i+1}-1}, S_i, S_{i+1}, x_{S_i}, x_{S_{i+1}}) - \sum_{t=S_i}^{S_{i+1}-1} u_t(x_t, x_{t+1}) > \gamma\}. \tag{2.126}$$

By (2.125) and (2.126),

$$M \ge \gamma \mathrm{Card}(E)$$

and

$$\mathrm{Card}(E) \le \gamma^{-1} M. \tag{2.127}$$

Let

$$j \in \{0, \dots, q-1\} \setminus E. \tag{2.128}$$

In view of (2.126) and (2.128),

$$\sum_{t=S_j}^{S_{j+1}-1} u_t(x_t, x_{t+1}) \geq \sigma(\{u_t\}_{t=S_j}^{S_{j+1}-1}, \{\Omega_t\}_{t=S_j}^{S_{j+1}-1}, S_j, S_{j+1}, x_{S_j}, x_{S_{j+1}}) - \gamma. \tag{2.129}$$

It follows from (2.116), (2.117), (2.123), (2.124), (2.129), and property (P11) that

$$\rho(x_t, \bar{x}) \leq \epsilon, \ t = S_j, \dots, S_{j+1}. \tag{2.130}$$

Since (2.130) holds for any integer j satisfying (2.128) we conclude that

$$\{t \in \{0, \dots, T\} : \ \rho(x_t, \bar{x}) > \epsilon\}$$

$$\subset \cup\{\{S_i, \dots, S_{i+1}\} : \ i \in E\} \cup \{0, \dots, S_0\} \cup \{S_q, \dots, T\}.$$

By the inclusion above, (2.115), (2.122), (2.123), and (2.127),

$$\mathrm{Card}(\{t \in \{0, \dots, T\} : \ \rho(x_t, \bar{x}) > \epsilon\})$$

$$\leq (L_0 + 1)(\mathrm{Card}(E) + 2) \leq (L_0 + 1)(2 + M\gamma^{-1}) < L.$$

This completes the proof of Theorems 2.4 and 2.5.

2.7 Proof of Theorem 2.6

By Lemma 2.16, there exists

$$\gamma \in (0, \min\{\epsilon, \bar{\lambda}\})$$

such that the following property holds:

(P12) For each natural number L there exists $\gamma_L \in (0, \gamma)$ such that for each natural number $T \leq L$, each

$$\Omega_t \in \mathcal{E}(\gamma_L), \ t = 0, \dots, T-1,$$

each $u_t \in \mathcal{M}, \ t = 0, \dots, T-1$ satisfying

$$\|u_t - v\| \leq \gamma_L, \ t = 0, \dots, T-1$$

and each $(\{\Omega_t\}_{t=0}^{T-1})$-program $\{y_t\}_{t=0}^T$ satisfying

$$\rho(y_0, \bar{x}), \ \rho(y_T, \bar{x}) \leq \gamma$$

and

$$\sum_{t=0}^{T-1} u_t(y_t, y_{t+1}) \geq \sigma(\{u_t\}_{t=0}^{T-1}, \{\Omega_t\}_{t=0}^{T-1}, 0, T, y_0, y_T) - \gamma$$

the inequality $\rho(y_t, \bar{x}) \leq \epsilon$ holds for all $t = 0, \dots, T$.

By Lemma 2.13 (with $\epsilon = \gamma$), there exists a natural number L_0 such that for each integer $L_1 \geq L_0$ there exists $\gamma_0 \in (0, \gamma)$ such that the following property holds:

(P13) For each integer $\tau \in [L_0, L_1]$, each

$$\Omega_t \in \mathcal{E}(\gamma_0), \ t = 0, \ldots, \tau - 1,$$

each $u_t \in \mathcal{M}$, $t = 0, \ldots, \tau - 1$ satisfying

$$\|u_t - v\| \leq \gamma_0, \ t = 0, \ldots, \tau - 1$$

and each $(\{\Omega_t\}_{t=0}^{\tau-1})$-program $\{x_t\}_{t=0}^{\tau}$ satisfying

$$\sum_{t=0}^{\tau-1} u_t(x_t, x_{t+1}) \geq \sum_{t=0}^{\tau-1} u_t(\bar{x}, \bar{x}) - M - 1$$

the inequality

$$\min\{\rho(x_t, \bar{x}) : \ t = 1, \ldots, \tau\} \leq \gamma$$

holds.

Let an integer $L_1 \geq L_0$ and let $\gamma_0 \in (0, \gamma)$ be as guaranteed by (P13).

By (P12) there exists $\delta \in (0, \gamma_0)$ such that the following property holds:

(P14) For each natural number $T \leq 2L_1 + 4$, each

$$\Omega_t \in \mathcal{E}(\delta), \ t = 0, \ldots, T - 1,$$

each $u_t \in \mathcal{M}$, $t = 0, \ldots, T - 1$ satisfying

$$\|u_t - v\| \leq \delta, \ t = 0, \ldots, T - 1$$

and each $(\{\Omega_t\}_{t=0}^{T-1})$-program $\{y_t\}_{t=0}^{T}$ satisfying

$$\rho(y_0, \bar{x}), \ \rho(y_T, \bar{x}) \leq \gamma$$

and

$$\sum_{t=0}^{T-1} u_t(y_t, y_{t+1}) \geq \sigma(\{u_t\}_{t=0}^{T-1}, \{\Omega_t\}_{t=0}^{T-1}, 0, T, y_0, y_T) - \gamma$$

the inequality $\rho(y_t, \bar{x}) \leq \epsilon$ holds for all $t = 0, \ldots, T$.

Assume that an integer $T > 3L_1$,

$$\Omega_t \in \mathcal{E}(\delta), \ t = 0, \ldots, T - 1, \tag{2.131}$$

$u_t \in \mathcal{M}$, $t = 0, \ldots, T - 1$ satisfy

$$\|u_t - v\| \leq \delta, \ t = 0, \ldots, T - 1 \tag{2.132}$$

and that an $(\{\Omega_t\}_{t=0}^{T-1})$-program $\{x_t\}_{t=0}^{T}$ and a finite sequence of integers $\{S_i\}_{i=0}^{q}$ satisfy

$$S_0 = 0, \ S_{i+1} - S_i \in [L_0, L_1], \ i = 0, \ldots, q - 1, \ S_q > T - L_1, \tag{2.133}$$

$$\sum_{t=S_i}^{S_{i+1}-1} u_t(x_t, x_{t+1}) \geq \sum_{t=S_i}^{S_{i+1}-1} u_t(\bar{x}, \bar{x}) - M \tag{2.134}$$

for each integer $i \in [0, q-1]$,

$$\sum_{t=S_i}^{S_{i+2}-1} u_t(x_t, x_{t+1}) \geq \sigma\left(\{u_t\}_{t=S_i}^{S_{i+2}-1}, \{\Omega_t\}_{t=S_i}^{S_{i+2}-1}, S_i, S_{i+2}, x_{S_i}, x_{S_{i+2}}\right) - \gamma \tag{2.135}$$

for each integer $i \in [0, q-2]$ and

$$\sum_{t=S_{q-2}}^{T-1} u_t(x_t, x_{t+1}) \geq \sigma(\{u_t\}_{t=S_{q-2}}^{T-1}, \{\Omega_t\}_{t=S_{q-2}}^{T-1}, S_{q-2}, T, x_{S_{q-2}}, x_T) - \gamma. \tag{2.136}$$

Let an integer $i \in [0, q-1]$. By (2.131)–(2.134), the inequality $\delta < \gamma_0$, the choice of γ_0 and property (P13), there exists an integer τ_i such that

$$\tau_i \in [S_i + 1, S_{i+1}] \text{ and } \rho(x_{\tau_i}, \bar{x}) \leq \gamma. \tag{2.137}$$

Thus for each integer $i \in [0, q-1]$ there exists an integer τ_i satisfying (2.137). Clearly,

$$\tau_0 \leq L_1, \ \tau_{q-1} > T - 2L_1.$$

Let an integer $i \in [0, q-2]$. By (2.133) and (2.137),

$$1 \leq \tau_{i+1} - \tau_i \leq 2L_1, \ \tau_i, \tau_{i+1} \in [S_i, S_{i+2}]. \tag{2.138}$$

By (2.136) and (2.138),

$$\sum_{t=\tau_i}^{\tau_{i+1}-1} u_t(x_t, x_{t+1}) \geq \sigma(\{u_t\}_{t=\tau_i}^{\tau_{i+1}-1}, \{\Omega_t\}_{\tau_i}^{\tau_{i+1}-1}, \tau_i, \tau_{i+1}, x_{\tau_i}, x_{\tau_{i+1}}) - \gamma. \tag{2.139}$$

Thus we have shown that there exists a finite sequence of integers $\{\tau_i\}_{i=0}^{p}$ such that

$$\rho(x_{\tau_i}, \bar{x}) \leq \gamma, i = 0, \dots, p,$$

$$1 \leq \tau_{i+1} - \tau_i \leq 2L_1 \text{ for all integers } i \text{ satisfying } 0 \leq i < p \tag{2.140}$$

and (2.139) holds for all integers i satisfying $0 \leq i < p$.

Clearly, we may assume that $\tau_0 = 0$ if $\rho(x_0, \bar{x}) \leq \gamma$ and $\tau_p = T$ if $\rho(x_T, \bar{x}) \leq \gamma$. Let an integer i satisfies $0 \leq i < p$. By (2.131), (2.132), (2.137), (2.139), (2.140), and property (P14),

$$\rho(x_t, \bar{x}) \leq \epsilon, \ t = \tau_i, \dots, \tau_{i+1}.$$

This implies that

$$\rho(x_t, \bar{x}) \leq \epsilon, \ t = \tau_0, \dots, \tau_p$$

and completes the proof of Theorem 2.6.

2.8   Proof of Theorems 2.7 and 2.8

We prove Theorems 2.7 and 2.8 simultaneously. In the case of Theorem 2.8 set $l_2 = 1$. We may assume without loss of generality that $\epsilon < \bar{\lambda}$.

By Theorem 2.2, there exist $\delta_1 \in (0, \epsilon)$ and a natural number $L_1 > l_1 + l_2$ such that the following property holds:

(P15) For each integer $T > 2L_1$, each

$$\Omega_t \in \mathcal{E}(\delta_1), \; t = 0, \ldots, T-1,$$

each $u_t \in \mathcal{M}$, $t = 0, \ldots, T-1$ satisfying

$$\|u_t - v\| \leq \delta_1, \; t = 0, \ldots, T-1$$

and each $(\{\Omega_t\}_{t=0}^{T-1})$-program $\{x_t\}_{t=0}^{T}$ which satisfies

$$\rho(x_0, \bar{x}) \leq \delta_1, \; \rho(x_T, \bar{x}) \leq \delta_1,$$

$$\sigma(\{u_t\}_{t=0}^{T-1}, \{\Omega_t\}_{t=0}^{T-1}, 0, T, x_0, x_T) \leq \sum_{t=0}^{T-1} u_t(x_t, x_{t+1}) + \delta_1$$

we have

$$\rho(x_t, \bar{x}) \leq \epsilon \text{ for all } t = 0, \ldots, T.$$

By Theorems 2.4 and 2.5, there exist a natural number

$$L_2 > 2L_1 \text{ and } \delta \in (0, \delta_1) \tag{2.141}$$

such that the following property holds:

(P16) For each integer $T > L_2$, each

$$\Omega_t \in \mathcal{E}(\delta), \; t = 0, \ldots, T-1,$$

each $u_t \in \mathcal{M}$, $t = 0, \ldots, T-1$ satisfying

$$\|u_t - v\| \leq \delta, \; t = 0, \ldots, T-1$$

and each $(\{\Omega_t\}_{t=0}^{T-1})$-program $\{x_t\}_{t=0}^{T}$ which satisfies at least one of the following conditions below:

(i)

$$x_0 \in \bar{Y}(\{\Omega_t\}_{t=0}^{l_1-1}, 0, l_1), \; x_T \in Y(\{\Omega_t\}_{t=T-l_2}^{T-1}, T-l_2, T),$$

$$\sigma(\{u_t\}_{t=0}^{T-1}, \{\Omega_t\}_{t=0}^{T-1}, 0, T, x_0, x_T) \leq \sum_{t=0}^{T-1} u_t(x_t, x_{t+1}) + M;$$

(ii)

$$x_0 \in \bar{Y}(\{\Omega_t\}_{t=0}^{l_1-1}, 0, l_1),$$

$$\sigma(\{u_t\}_{t=0}^{T-1}, \{\Omega_t\}_{t=0}^{T-1}, 0, T, x_0) \leq \sum_{t=0}^{T-1} u_t(x_t, x_{t+1}) + M$$

the inequality

$$Card(\{t \in \{0, \ldots, T\} :\ \rho(x_t, \bar{x}) > \delta_1\}) \leq L_2$$

holds.

Set

$$L = 4L_2.$$

Assume that an integer $T > 2\,L$,

$$\Omega_t \in \mathcal{E}(\delta),\ t = 0, \ldots, T - 1, \tag{2.142}$$

$u_t \in \mathcal{M},\ t = 0, \ldots, T - 1$ satisfies

$$\|u_t - v\| \leq \delta,\ t = 0, \ldots, T - 1 \tag{2.143}$$

and that an $(\{\Omega_t\}_{t=0}^{T-1})$-program $\{x_t\}_{t=0}^{T}$ satisfies

$$\sum_{t=\tau}^{\tau+L-1} u_t(x_t, x_{t+1}) \geq \sigma(\{u_t\}_{t=\tau}^{\tau+L-1}, \{\Omega_t\}_{t=\tau}^{\tau+L-1}, \tau, \tau + L, x_\tau, x_{\tau+L}) - \delta \tag{2.144}$$

for each integer $\tau \in [0, T - L]$ and at least one of the conditions (i) and (ii) hold.

Set

$$E_0 = \{t \in \{0, \ldots, T\} :\ \rho(x_t, \bar{x}) > \delta_1\}. \tag{2.145}$$

In view of (2.142), (2.143), (2.145), (P16), and conditions (i) and (ii),

$$Card(E_0) \leq L_2. \tag{2.146}$$

Set

$$\tau_1 = \min\{t \in \{0, \ldots, T\} :\ \rho(x_t, \bar{x}) \leq \delta_1\}, \tag{2.147}$$

$$\tau_2 = \max\{t \in \{0, \ldots, T\} :\ \rho(x_t, \bar{x}) \leq \delta_1\}, \tag{2.148}$$

By (2.145) and (2.148),

$$\tau_1 \leq L_2,\ \tau_2 \geq T - L_2. \tag{2.149}$$

In order to complete the proof of Theorems 2.7 and 2.8 it is sufficient to show that

$$\rho(x_t, \bar{x}) \leq \epsilon \text{ for all } t = \tau_1, \ldots, \tau_2.$$

Let an integer

$$t \in [\tau_1, \tau_2]. \tag{2.150}$$

In view of (2.149), (2.150), and the inequality $T > 2\,L$, at least one of the following inequalities holds:

$$t - \tau_1 > 2L_2;\ \tau_2 - t > 2L_2. \tag{2.151}$$

Define integers $\tilde{\tau}_1, \tilde{\tau}_2$ as follows:

$$\text{if} t - \tau_1 \leq 2L_2 \text{ set} \tilde{\tau}_1 = \tau_1; \tag{2.152}$$

if $t - \tau_1 > 2L_2$, then by (2.145) and (2.146) there exists an integer $\tilde{\tau}_1$ such that

$$\tilde{\tau}_1 \in [t - 2L_2, t - L_2], \ \rho(x_{\tilde{\tau}_1}, \bar{x}) \leq \delta_1; \tag{2.153}$$

$$\text{if } \tau_2 - t \leq 2L_2 \text{ set } \tilde{\tau}_2 = \tau_2; \tag{2.154}$$

if $\tau_2 - t > 2L_2$, then by (2.145) and (2.146) there exists an integer $\tilde{\tau}_2$ such that

$$\tilde{\tau}_2 \in [t + L_2, t + 2L_2], \ \rho(x_{\tilde{\tau}_2}, \bar{x}) \leq \delta_1. \tag{2.155}$$

By (2.151)–(2.155),

$$\tilde{\tau}_2 - \tilde{\tau}_1 = \tilde{\tau}_2 - t + t - \tilde{\tau}_1 \leq 4L_2 = L, \tag{2.156}$$

$$\tilde{\tau}_2 - \tilde{\tau}_1 = \tilde{\tau}_2 - t + t - \tilde{\tau}_1 \geq \max\{\tilde{\tau}_2 - t, \ t - \tilde{\tau}_1\} \geq L_2. \tag{2.157}$$

It follows from (2.144), (2.156), and (2.157) that

$$\sum_{t=\tilde{\tau}_1}^{\tilde{\tau}_2-1} u_t(x_t, x_{t+1}) \geq \sigma(\{u_t\}_{t=\tilde{\tau}_1}^{\tilde{\tau}_2-1}, \{\Omega_t\}_{t=\tilde{\tau}_1}^{\tilde{\tau}_2-1}, \tilde{\tau}_1, \tilde{\tau}_2, x_{\tilde{\tau}_1}, x_{\tilde{\tau}_2}) - \delta. \tag{2.158}$$

In view of (2.141)–(2.143), (2.147), (2.148), (2.152)–(2.158), and property (P15),

$$\rho(x_t, \bar{x}) \leq \epsilon.$$

This completes the proof of Theorems 2.7 and 2.8.

Chapter 3
Optimal Control Problems with Discounting

In this chapter we continue our study of the structure of approximate solutions of the discrete-time optimal control problems with a compact metric space of states X and with a singleton turnpike. These problems are described by a nonempty closed set $\Omega \subset X \times X$ which determines a class of admissible trajectories (programs) and by a bounded upper semicontinuous objective function $v : X \times X \to R^1$ which determines an optimality criterion. We show the stability of the turnpike phenomenon under small perturbations of the objective function v and the set Ω in the case with discounting. The results of the chapter generalize the results obtained in [54] for the discounting case with a perturbation only on the objective function.

3.1 Stability of the Turnpike Phenomenon

We use the notation, definitions, and assumptions introduced in Sect. 2.1. Denote by $\mathcal{M}_0$ the set of all upper semicontinuous functions $u \in \mathcal{M}$.

It is not difficult to see that the following result holds.

Proposition 3.1 *Let l be a natural number, integers T_1, T_2 satisfy $0 \leq T_1 \leq T_2 - l$, $\{u_t\}_{t=T_1}^{T_2-1} \subset \mathcal{M}_0$, for any integer $t \in \{T_1, \ldots, T_2 - 1\}$, let Ω_t be a closed subset of $X \times X$ such that $(\bar{x}, \bar{x}) \in \Omega_t$ and let*

$$x \in \bar{Y}\left(\{\Omega_t\}_{t=T_1}^{T_1+l-1}, 0, l\right).$$

Then there exists an $(\{\Omega_t\}_{t=T_1}^{T_2-1})$-program $\{x_t\}_{t=T_1}^{T_2}$ such that $x_0 = x$ and

$$\sum_{t=T_1}^{T_2-1} u_t(x_t, x_{t+1}) = \sigma\left(\{u_t\}_{t=T_1}^{T_2-1}, \{\Omega_t\}_{t=T_1}^{T_2-1}, T_1, T_2, x\right).$$

In this chapter we prove the following result which shows the stability of the turnpike phenomenon in the case of discounting.

A. J. Zaslavski, *Stability of the Turnpike Phenomenon in Discrete-Time Optimal Control Problems*, SpringerBriefs in Optimization, DOI 10.1007/978-3-319-08034-5_3, © The Author 2014

Theorem 3.2 *Let $\epsilon \in \left(0, \bar{\lambda}\right)$ and let l be a natural number. Then there exist $\delta \in (0, \epsilon)$, a natural number $L > l$ and $\lambda \in (0, 1)$ such that for each integer $T > 2L$, each*

$$\Omega_t \in \mathcal{E}(\delta), \ t = 0, \ldots, T - 1,$$

each $u_t \in \mathcal{M}_0$, $t = 0, \ldots, T - 1$ satisfying

$$\|u_t - v\| \le \delta, \ t = 0, \ldots, T - 1,$$

each sequence $\{\alpha_t\}_{t=0}^{T-1} \subset (0, 1]$ such that

$$\alpha_i \alpha_j^{-1} \ge \lambda \text{ for each } i, j \in \{0, \ldots, T - 1\} \text{ satisfying } |i - j| \le L,$$

and each $\left(\{\Omega_t\}_{t=0}^{T-1}\right)$-program $\{x_t\}_{t=0}^{T}$ which satisfies

$$x_0 \in \bar{Y}\left(\{\Omega_t\}_{t=0}^{l-1}, 0, l\right),$$

$$\sigma\left(\{\alpha_t u_t\}_{t=0}^{T-1}, \{\Omega_t\}_{t=0}^{T-1}, 0, T, x_0\right) = \sum_{t=0}^{T-1} \alpha_t u_t (x_t, x_{t+1})$$

there exist integers $\tau_1 \in [0, L]$, $\tau_2 \in [T - L, T]$ such that

$$\rho(x_t, \bar{x}) \le \epsilon \text{ for all } t = \tau_1, \ldots, \tau_2.$$

Moreover if $\rho(x_0, \bar{x}) \le \delta$, then $\tau_1 = 0$ and if $\rho(x_T, \bar{x}) \le \delta$, then $\tau_2 = T$.

Roughly speaking, the turnpike property holds if discount coefficients $\{\alpha_t\}_{t=0}^{T-1} \subset (0, 1]$ are changed rather slowly.

Let $\Omega_t \subset X \times X$ be a nonempty set for all nonnegative integers t.

A sequence $\{x_t\}_{t=0}^{\infty} \subset X$ is called an $\left(\{\Omega_t\}_{t=0}^{\infty}\right)$-program if $(x_t, x_{t+1}) \in \Omega_t$ for all integers $t \ge 0$.

Let $\{u_t\}_{t=0}^{\infty} \subset \mathcal{M}$ be given. An $\left(\{\Omega_t\}_{t=0}^{\infty}\right)$-program $\{x_t\}_{t=0}^{\infty}$ is called $\left(\{u_t\}_{t=0}^{\infty}, \{\Omega_t\}_{t=0}^{\infty}\right)$-overtaking optimal if for each $\left(\{\Omega_t\}_{t=0}^{\infty}\right)$-program $\{y_t\}_{t=0}^{\infty}$ satisfying $x_0 = y_0$, we have

$$\limsup_{T \to \infty} \left[\sum_{t=0}^{T-1} u_t (y_t, y_{t+1}) - \sum_{t=0}^{T-1} u_t (x_t, x_{t+1})\right] \le 0.$$

The following result establishes the stability of turnpike phenomenon for overtaking optimal programs.

Theorem 3.3 *Let $\epsilon \in \left(0, \bar{\lambda}\right)$ and let l be a natural number. Then there exist $\delta \in (0, \epsilon)$, a natural number $L > l$ and $\lambda \in (0, 1)$ such that for each*

$$\Omega_t \in \mathcal{E}(\delta), \ t = 0, 1, \ldots,$$

each $u_t \in \mathcal{M}_0$, $t = 0, 1, \ldots$, satisfying

$$\|u_t - v\| \le \delta, \ t = 0, 1, \ldots,$$

each sequence $\{\alpha_t\}_{t=0}^{\infty} \subset (0,1]$ such that

$$\alpha_i \alpha_j^{-1} \geq \lambda \text{ for each pair of nonnegative integers } i, j \text{ satisfying } |i-j| \leq L$$

and each $\left(\{\alpha_t u_t\}_{t=0}^{\infty}, \{\Omega_t\}_{t=0}^{\infty}\right)$-overtaking optimal program $\{x_t\}_{t=0}^{\infty}$ which satisfies

$$x_0 \in \bar{Y}\left(\{\Omega_t\}_{t=0}^{l-1}, 0, l\right)$$

the following inequality holds:

$$\rho(x_t, \bar{x}) \leq \epsilon \text{ for all integers } t \geq L.$$

Moreover, if $\rho(x_0, \bar{x}) \leq \delta$, then

$$\rho(x_t, \bar{x}) \leq \epsilon \text{ for all integers } t \geq 0.$$

In this chapter we prove the following existence result.

Theorem 3.4 *Let $l \geq 1$ be an integer, $\epsilon = \bar{\lambda}/4$, and let $\delta \in (0, \epsilon)$, an integer $L > l$ and $\lambda \in (0, 1)$ be as guaranteed by Theorem 3.2.*
    Let

$$u_t \in \mathcal{M}_0 \text{ and } \|u_t - v\| \leq \delta, \ t = 0, 1, \ldots,$$
$$\Omega_t \in \mathcal{E}(\delta), \ t = 0, 1, \ldots$$

be closed subsets of $X \times X$ for all nonnegative integers t, and let $\{\alpha_t\}_{t=0}^{\infty} \subset (0,1]$ satisfy the relations
$$\lim_{t \to \infty} \alpha_t = 0,$$

$$\alpha_i \alpha_j^{-1} \geq \lambda \text{ for all nonnegative integers } i, j \text{ satisfying } |i-j| \leq L.$$

Then for each $z \in \bar{Y}\left(\{\Omega_t\}_{t=0}^{l-1}, 0, l\right)$ there exists an $\left(\{\Omega_t\}_{t=0}^{\infty}\right)$-program $\left\{x_t^{(z)}\right\}_{t=0}^{\infty}$ such that $x_0^{(z)} = z$ and that the following property holds:
    For each real number $\gamma > 0$ there exists an integer $n_0 \geq 1$ such that for each integer $T \geq n_0$ and each point $z \in \bar{Y}\left(\{\Omega_t\}_{t=0}^{l-1}, 0, l\right)$ the inequality

$$\left| \sigma\left(\{\alpha_t u_t\}_{t=0}^{T-1}, \{\Omega_t\}_{t=0}^{T-1}, 0, T, z\right) - \sum_{t=0}^{T-1} \alpha_t u_t\left(x_t^{(z)}, x_{t+1}^{(z)}\right) \right| \leq \gamma$$

holds.

It is clear that Theorem 3.4 establishes the existence of $\left(\{\alpha_t u_t\}_{t=0}^{\infty}\right)$-overtaking optimal program when the sequence of the discount coefficients $\{\alpha_t\}_{t=0}^{\infty}$ tends to zero slowly.

Note that the existence of an $\left(\{\alpha_t u_t\}_{t=0}^{\infty}\right)$-overtaking optimal program when the discount coefficients $\{\alpha_t\}_{t=0}^{\infty}$ tends to zero rapidly is a well known fact.

3.2   Auxiliary Results

In the proof of Theorems 3.2 and 3.3 we use the following two lemmas.

Lemma 3.5 *Let $\epsilon \in (0, \bar{\lambda})$ and l be a natural number. Then there exist $\delta \in (0, \epsilon)$, a natural number $L > l$ and $\lambda \in (0, 1)$ such that for each integer $T \geq L$, each*

$$\Omega_t \in \mathcal{E}(\delta), \ t = 0, \ldots, T - 1,$$

each $u_t \in \mathcal{M}, \ t = 0, \ldots, T - 1$ satisfying

$$\|u_t - v\| \leq \delta, \ t = 0, \ldots, T - 1,$$

each finite sequence $\{\alpha_t\}_{t=0}^{T-1} \subset (0, 1]$ which satisfies

$$\alpha_i \alpha_j^{-1} \geq \lambda \text{ for each } i, j \in \{0, 1, \ldots, T - 1\} \text{ satisfying } |i - j| \leq L,$$

each $\left(\{\Omega_t\}_{t=0}^{T-1}\right)$-program $\{x_t\}_{t=0}^{T}$ such that

$$\sum_{t=0}^{T-1} \alpha_t u_t(x_t, x_{t+1}) = \sigma \left(\{\alpha_t u_t\}_{t=0}^{T-1}, \{\Omega_t\}_{t=0}^{T-1}, 0, T, x_0\right)$$

and each integer $S \in [0, T - L]$ satisfying

$$x_S \in \bar{Y}\left(\{\Omega_t\}_{t=S}^{S+l-1}, S, S + l\right) \tag{3.1}$$

the inequality

$$\min\{\rho(x_t, \bar{x}) : \ t = S + 1, \ldots, S + L\} \leq \epsilon \tag{3.2}$$

holds.

Lemma 3.6 *Let $\epsilon \in (0, \bar{\lambda})$ and l be a natural number. Then there exist $\delta \in (0, \epsilon)$, a natural number L and $\lambda \in (0, 1)$ such that for each*

$$\Omega_t \in \mathcal{E}(\delta), \ t = 0, 1, \ldots,$$

each $u_t \in \mathcal{M}, \ t = 0, 1, \ldots,$ satisfying

$$\|u_t - v\| \leq \delta, \ t = 0, 1, \ldots,$$

each sequence $\{\alpha_t\}_{t=0}^{\infty} \subset (0, 1]$ which satisfies

$$\alpha_i \alpha_j^{-1} \geq \lambda \text{ for each pair of integers } i, j \geq 0 \text{ satisfying } |i - j| \leq L,$$

each $\left(\{\alpha_t u_t\}_{t=0}^{\infty}, \{\Omega_t\}_{t=0}^{\infty}\right)$-overtaking optimal program $\{x_t\}_{t=0}^{\infty}$ and each integer $S \geq 0$ satisfying (3.1), inequality (3.2) holds.

Proof We prove Lemmas 3.5 and 3.6 simultaneously. By Lemma 2.13, there exist a natural number L_0 and $\delta \in (0, \epsilon)$ such that the following property holds:

(P17) For each

$$\Omega_t \in \mathcal{E}(\delta), \ t = 0, \ldots, L_0 - 1,$$

each $u_t \in \mathcal{M}, \ t = 0, \ldots, L_0 - 1$ satisfying

$$\|u_t - v\| \leq \delta, \ t = 0, \ldots, L_0 - 1$$

and each $(\{\Omega_t\}_{t=0}^{L_0-1})$-program $\{x_t\}_{t=0}^{L_0}$ satisfying

$$\sum_{t=0}^{L_0-1} u_t(x_t, x_{t+1}) \geq \sum_{t=0}^{L_0-1} u_t(\bar{x}, \bar{x}) - 1$$

the inequality

$$\min\{\rho(x_t, \bar{x}) : \ t = 1, \ldots, L_0\} \leq \epsilon$$

holds.

Choose a natural number k_0 such that

$$k_0 > 8(\|v\| + 1)(L_0 + l + 1). \tag{3.3}$$

Set

$$L = L_0 k_0. \tag{3.4}$$

Choose a number $\lambda \in (0, 1)$ such that

$$\lambda^l > 2^{-1}, \tag{3.5}$$

$$8(\|v\| + 1)L_0|1 - \lambda| < 2^{-1}, \tag{3.6}$$

$$\lambda^{k_0+l+1} k_0 > 8(\|v\| + 1)(L_0 + l + 1). \tag{3.7}$$

We suppose that

$$\infty + x = \infty \text{ for any } x \in R^1$$

and that the sum over empty set is zero.

Let

$$T \in \{1, 2 \ldots, \} \cup \{\infty\} \text{ and } T \geq L, \tag{3.8}$$

for all integers t satisfying $0 \leq t < T$,

.

$$\Omega_t \in \mathcal{E}(\delta), \tag{3.9}$$

$$u_t \in \mathcal{M} \text{ and } \|u_t - v\| \leq \delta, \tag{3.10}$$

$$\alpha_t \in (0, 1] \tag{3.11}$$

satisfy

$$\alpha_i \alpha_j^{-1} \geq \lambda \text{ for each pair of integers } i, j \geq 0 \text{ satisfying } i, j < T \text{ and } |i - j| \leq L, \tag{3.12}$$

and let an $\left(\{\Omega_t\}_{t=0}^{T-1}\right)$-program $\{x_t\}_{t=0}^{T}$ satisfy

$$\sum_{t=0}^{T-1} \alpha_t u_t (x_t, x_{t+1}) = \sigma \left(\{\alpha_t u_t\}_{t=0}^{T-1}, \{\Omega_t\}_{t=0}^{T-1}, 0, T, x_0\right) \tag{3.13}$$

in the case of Lemma 3.5, and let

$$\{x_t\}_{t=0}^{\infty} \text{ be an } \left(\{\alpha_t u_t\}_{t=0}^{\infty}, \{\Omega_t\}_{t=0}^{\infty}\right) \text{-overtaking optimal program} \tag{3.14}$$

in the case of Lemma 3.6.

Assume that an integer $S \geq 0$ satisfies

$$S \leq T - L, \ x_S \in \bar{Y} \left(\{\Omega_t\}_{t=S}^{S+l-1}, S, S+l\right). \tag{3.15}$$

In order to complete the proof of Lemmas 3.5 and 3.6, it is sufficient to show that

$$\min \{\rho(x_t, \bar{x}) : \ t = S + 1, \ldots, S + L\} \leq \epsilon.$$

Assume the contrary. Then

$$\rho(x_t, \bar{x}) > \epsilon \text{ for all } t = S + 1, \ldots, S + L. \tag{3.16}$$

There are two cases:

(1) There is an integer $S_0 > S$ such that $S \leq T$ and

$$\rho \left(x_{S_0}, \bar{x}\right) \leq \epsilon. \tag{3.17}$$

(2)

$$\rho(x_t, \bar{x}) > \epsilon \text{ for all integers } t \text{ satisfying } S + 1 \leq t \leq T. \tag{3.18}$$

Assume that the case (1) holds. Thus (3.17) holds. In view of (3.16) and (3.17),

$$S_0 > S + L. \tag{3.19}$$

We may assume without loss of generality that

$$\rho \left(x_t, \bar{x}\right) > \epsilon \text{ for all } t = S + 1, \ldots, S_0 - 1. \tag{3.20}$$

In view of (3.15), there exists an $\left(\{\Omega_t\}_{t=S}^{S+l-1}\right)$-program $\{y_t\}_{t=S}^{S+l}$ such that

$$y_S = x_S, \quad y_{S+l} = \bar{x}. \tag{3.21}$$

Set

$$y_{S_0} = x_{S_0}, \quad y_t = \bar{x}, \quad t = S + l + 1, \ldots, S_0 - 1. \tag{3.22}$$

By (2.4), (3.17), (3.21), (3.22), and the inequality $\epsilon < \bar{\lambda}$, $\{y_t\}_{t=S}^{S_0}$ is an $\{\Omega_t\}_{t=S}^{S_0-1}$-program. It follows from (3.10), (3.12), (3.13), (3.14), (3.21), and (3.22) that

$$\sum_{t=S}^{S_0-1} \alpha_t u_t\,(x_t, x_{t+1}) \geq \sum_{t=S}^{S_0-1} \alpha_t u_t\,(y_t, y_{t+1})$$

$$\geq \sum_{t=S}^{S_0-1} \alpha_t u_t\,(\bar{x}, \bar{x}) - 2 \sum_{t=S}^{S+l-1} \alpha_t \|u_t\| - 2\|u_{S_0-1}\| \alpha_{S_0-1}$$

$$\geq \sum_{t=S}^{S_0-1} \alpha_t u_t\,(\bar{x}, \bar{x}) - 2\alpha_S \lambda^{-1} l(\|v\| + 1) - 2\alpha_{S_0-1}(\|v\| + 1). \tag{3.23}$$

There exists a natural number k such that

$$S_0 - S \in (kL_0, (k+1)L_0]. \tag{3.24}$$

By (3.4), (3.19), and (3.24),

$$k \geq k_0.$$

It follows from (3.9), (3.10), (3.20), (3.24), and property (P17) that for each integer $j \in [0, k-1]$,

$$\sum_{t=S+jL_0}^{S+(j+1)L_0-1} u_t\,(x_t, x_{t+1}) < \sum_{t=S+jL_0}^{S+(j+1)L_0-1} u_t(\bar{x}, \bar{x}) - 1. \tag{3.25}$$

Let an integer

$$j \in [0, k-1]. \tag{3.26}$$

By (3.5), (3.6), (3.10), (3.25), and (3.26),

$$\sum_{t=S+jL_0}^{S+(j+1)L_0-1} \alpha_t u_t\,(x_t, x_{t+1}) - \sum_{t=S+jL_0}^{S+(j+1)L_0-1} \alpha_t u_t(\bar{x}, \bar{x})$$

$$= \alpha_{S+jL_0} \left[\sum_{t=S+jL_0}^{S+(j+1)L_0-1} u_t\,(x_t, x_{t+1}) - \sum_{t=S+jL_0}^{S+(j+1)L_0-1} u_t(\bar{x}, \bar{x}) \right]$$

$$+\alpha_{S+jL_0}\left[\alpha_{S+jL_0}^{-1}\sum_{t=S+jL_0}^{S+(j+1)L_0-1}\alpha_t u_t\,(x_t,x_{t+1})-\sum_{t=S+jL_0}^{S+(j+1)L_0-1}u_t(x_t,x_{t+1})\right]$$

$$+\alpha_{S+jL_0}\left[\sum_{t=S+jL_0}^{S+(j+1)L_0-1}u_t(\bar{x},\bar{x})-\alpha_{S+jL_0}^{-1}\sum_{t=S+jL_0}^{S+(j+1)L_0-1}\alpha_t u_t(\bar{x},\bar{x})\right]$$

$$<-\alpha_{S+jL_0}+\alpha_{S+jL_0}\left[\sum_{t=S+jL_0}^{S+(j+1)L_0-1}2|\alpha_t\alpha_{S+jL_0}^{-1}-1|(\|v\|+1)\right]$$

$$+\alpha_{S+jL_0}\left[\sum_{t=S+jL_0}^{S+(j+1)L_0-1}2|\alpha_t\alpha_{S+jL_0}^{-1}-1|(\|v\|+1)\right]$$

$$\leq\alpha_{S+jL_0}\left[-1+4(\|v\|+1)L_0|1-\lambda|\lambda^{-1}|\right]$$

$$\leq\alpha_{S+jL_0}\left[-1+8(\|v\|+1)L_0|1-\lambda|\right]<-2^{-1}\alpha_{S+jL_0}. \qquad (3.27)$$

By (3.10), (3.12), (3.23), (3.24), and (3.27),

$$-2\alpha_S\lambda^{-1}l(\|v\|+1)-2\alpha_{S_0-1}(\|v\|+1)$$

$$\leq\sum_{t=S}^{S_0-1}\alpha_t u_t\,(x_t,x_{t+1})-\sum_{t=S}^{S_0-1}\alpha_t u_t(\bar{x},\bar{x})$$

$$=\sum_{j=0}^{k-1}\left[\sum_{t=S+jL_0}^{S+(j+1)L_0-1}\alpha_t u_t\,(x_t,x_{t+1})-\sum_{t=S+jL_0}^{S+(j+1)L_0-1}\alpha_t u_t(\bar{x},\bar{x})\right]$$

$$+2(\|v\|+1)L_0\max\{\alpha_t:\,t=S+kL_0,\ldots,S_0\}$$

$$\leq-2^{-1}\sum_{j=0}^{k-1}\alpha_{S+jL_0}+2(\|v\|+1)L_0\lambda^{-1}\alpha_{S+kL_0}.$$

It follows from the relation above, (3.12), and (3.24) that

$$\sum_{j=0}^{k-1}\alpha_{S+jL_0}\leq4(\|v\|+1)\left[\alpha_S\lambda^{-1}l+\alpha_{S_0-1}+L_0\lambda^{-1}\alpha_{S+kL_0}\right]$$

$$\leq4(\|v\|+1)\left[\alpha_S\lambda^{-1}l+(L_0+1)\lambda^{-1}\alpha_{S+kL_0}\right].$$

In view of the inequality above, (3.5), (3.7), (3.12), and (3.25),

$$0\geq\sum_{j=0}^{k-1}\alpha_{S+jL_0}-4(\|v\|+1)\left[\alpha_S\lambda^{-1}l+(L_0+1)\lambda^{-1}\alpha_{S+kL_0}\right]$$

$$=2^{-1}\sum_{j=0}^{k-1}\alpha_{S+jL_0}-4(\|v\|+1)\alpha_S\lambda^{-1}l$$

$$+ 2^{-1} \sum_{j=0}^{k-1} \alpha_{S+jL_0} - 4(\|v\| + 1)(L_0 + 1)\lambda^{-1} \alpha_{S+kL_0}$$

$$\geq 2^{-1} \sum_{j=0}^{k_0-1} \alpha_S \lambda^j - 4(\|v\| + 1)\alpha_S \lambda^{-1} l$$

$$+ 2^{-1} \alpha_{S+kL_0} \sum_{j=k-k_0}^{k-1} \lambda^{k-j} - 4(\|v\| + 1)(L_0 + 1)\lambda^{-1} \alpha_{S+kL_0}$$

$$\geq \alpha_S [2^{-1} k_0 \lambda^{k_0} - 4(\|v\| + 1)\lambda^{-1} l]$$

$$+ \alpha_{S+kL_0} \left[2^{-1} k_0 \lambda^{k_0} - 4(\|v\| + 1)(L_0 + 1)\lambda^{-1} \right] > 0,$$

a contradiction. The contradiction we have reached proves that case (1) does not hold. Therefore, case (2) holds and (3.18) is true.

In view of (3.15), there exists an $\left(\{\Omega_t\}_{t=S}^{S+l-1}\right)$-program $\{y_t\}_{t=S}^{S+l}$ such that

$$y_S = x_S, \ y_{S+l} = \bar{x}. \tag{3.28}$$

Set

$$y_t = \bar{x} \text{ for all integers } t \text{ satisfying } S + l < t \leq T. \tag{3.29}$$

By (3.28) and (3.29), $\{y_t\}_{t=S}^{T}$ is an $\left(\{\Omega_t\}_{t=S}^{T-1}\right)$-program.

It follows from (3.9), (3.10), (3.18), and property (P17) that that for each integer $j \geq 0$ satisfying $S + (j + 1)L_0 \leq T$,

$$\sum_{t=S+jL_0}^{S+(j+1)L_0-1} u_t(x_t, x_{t+1}) < \sum_{t=S+jL_0}^{S+(j+1)L_0-1} u_t(\bar{x}, \bar{x}) - 1.$$

Arguing as in the case (1) we show that for each integer $j \geq 0$ satisfying $S + (j + 1)L_0 \leq T$, (3.27) holds.

Consider the case of Lemma 3.6. By (3.7), (3.10), (3.12), (3.14), (3.28), and (3.27) which holds for each integer $j \geq 0$,

$$0 \geq \limsup_{k\to\infty} \left(\sum_{t=S}^{S+kL_0-1} \alpha_t u_t (y_t, y_{t+1}) - \sum_{t=S}^{S+kL_0-1} \alpha_t u_t(x_t, x_t) \right)$$

$$= \limsup_{k\to\infty} \left(\sum_{t=S}^{S+kL_0-1} \alpha_t u_t (\bar{x}, \bar{x}) - \sum_{t=S}^{S+kL_0-1} \alpha_t u_t(x_t, x_t) \right)$$

$$+ \sum_{t=S}^{S+l-1} \alpha_t u_t (y_t, y_{t+1}) - \sum_{t=S}^{S+l-1} \alpha_t u_t(\bar{x}, \bar{x})$$

$$\geq \limsup_{k\to\infty} \left[\sum_{j=0}^{k-1} \left(\sum_{t=S+jL_0}^{S+(j+1)L_0-1} \alpha_t u_t(\bar{x}, \bar{x}) - \sum_{t=S+jL_0}^{S+(j+1)L_0-1} \alpha_t u_t(x_t, x_t) \right) \right]$$

$$- 2(\|v\| + 1)l\alpha_S \lambda^{-1}$$

$$\geq \limsup_{k \to \infty} \sum_{j=0}^{k-1} 2^{-1} \alpha_{S+jL_0} - 2(\|v\| + 1)l\alpha_S \lambda^{-1}$$

$$\geq \sum_{j=0}^{k_0-1} \alpha_S \lambda^j - 2(\|v\| + 1)l\alpha_S \lambda^{-1}$$

$$\geq \alpha_S \left(k_0 \lambda^{k_0} - 2(\|v\| + 1)l\lambda^{-1} \right) > 0,$$

a contradiction. The contradiction we have reached proves Lemma 3.6.

Let us complete the proof of Lemma 3.5. There exists a natural number k such that

$$T - S \in [kL_0, (k+1)L_0). \tag{3.30}$$

By (3.4), (3.15), and (3.30),

$$k \geq k_0.$$

By the inequality above, (3.7), (3.10), (3.12), (3.13), (3.28)–(3.30), and (3.27) which holds for each integer $j \geq 0$ satisfying $S + (j+1)L_0 \leq T$,

$$0 \geq \sum_{t=S}^{T-1} \alpha_t u_t \left(y_t, y_{t+1} \right) - \sum_{t=S}^{T-1} \alpha_t u_t \left(x_t, x_t \right)$$

$$\geq \sum_{t=S}^{T-1} \alpha_t u_t(\bar{x}, \bar{x}) - \sum_{t=S}^{T-1} \alpha_t u_t(x_t, x_t)$$

$$+ \sum_{t=S}^{S+l-1} \alpha_t u_t \left(y_t, y_{t+1} \right) - \sum_{t=S}^{S+l-1} \alpha_t u_t(\bar{x}, \bar{x})$$

$$= \sum_{j=0}^{k-1} \left[\sum_{t=S+jL_0}^{S+(j+1)L_0-1} \alpha_t u_t(\bar{x}, \bar{x}) - \sum_{t=S+jL_0}^{S+(j+1)L_0-1} \alpha_t u_t(x_t, x_t) \right]$$

$$+ \sum \{\alpha_t u_t(\bar{x}, \bar{x}) - \alpha_t u_t(x_t, x_t) : \ t \text{ is an integer and } S + kL_0 \leq t < T\}$$

$$- 2(\|v\| + 1)l\alpha_S \lambda^{-l}$$

$$\geq 2^{-1} \sum_{j=0}^{k-1} \alpha_{S+jL_0} - 2L_0(\|v\| + 1)\alpha_{S+kL_0}\lambda^{-1} - 2(\|v\| + 1)\alpha_S \lambda^{-l}l$$

$$\geq 4^{-1}\alpha_S \sum_{j=0}^{k_0-1} \lambda^j - 2(\|v\| + 1)\alpha_S \lambda^{-l}l$$

$$+ 4^{-1} \sum_{j=k-k_0}^{k-1} \alpha_{S+kL_0} \lambda^{k-j} - 2L_0(\|v\| + 1)\alpha_{S+kL_0}\lambda^{-1}$$

$$\geq \alpha_S \left(4^{-1}k_0\lambda^{k_0} - 2(\|v\| + 1)\lambda^{-l}l\right)$$

$$+ \alpha_{S+kL_0} \left(4^{-1}k_0\lambda^{k_0} - 2L_0(\|v\| + 1)\lambda^{-1}\right) > 0,$$

a contradiction. The contradiction we have reached proves Lemma 3.5.

3.3 Proofs of Theorems 3.2 and 3.3

We prove Theorems 3.2 and 3.3 simultaneously.

By Lemma 2.16, there exists $\gamma \in (0, \min\{\bar{\lambda}, \epsilon\})$ such that the following property holds:

(P18) For each natural number L there exists $\gamma_L \in (0, \gamma)$ such that for each natural number $S \leq L$, each

$$\Omega_t \in \mathcal{E}(\gamma_L), \ t = 0, \ldots, S - 1,$$

each $u_t \in \mathcal{M}, \ t = 0, \ldots, S - 1$ satisfying

$$\|u_t - v\| \leq \gamma_L, \ t = 0, \ldots, S - 1$$

and each $\left(\{\Omega_t\}_{t=0}^{S-1}\right)$-program $\{y_t\}_{t=0}^{S}$ satisfying

$$\rho(y_0, \bar{x}), \ \rho(y_S, \bar{x}) \leq \gamma$$

and

$$\sum_{t=0}^{S-1} u_t(y_t, y_{t+1}) \geq \sigma \left(\{u_t\}_{t=0}^{S-1}, \{\Omega_t\}_{t=0}^{S-1}, 0, S, y_0, y_S\right) - \gamma$$

the inequality $\rho(y_t, \bar{x}) \leq \epsilon$ holds for all $t = 0, \ldots, S$.

By Lemmas 3.5 and 3.6 (with $\epsilon = \gamma$), there exist

$$\tilde{\gamma} \in (0, \gamma),$$

a natural number $L > l$ and $\tilde{\lambda} \in (0, 1)$ such that the following properties hold:

(i) For each integer $T \geq L$, each

$$\Omega_t \in \mathcal{E}(\tilde{\gamma}), \ t = 0, \ldots, T - 1,$$

each $u_t \in \mathcal{M}, \ t = 0, \ldots, T - 1$ satisfying

$$\|u_t - v\| \leq \tilde{\gamma}, \ t = 0, \ldots, T - 1,$$

each finite sequence $\{\alpha_t\}_{t=0}^{T-1} \subset (0, 1]$ which satisfies

$$\alpha_i \alpha_j^{-1} \geq \tilde{\lambda} \text{ for each } i, j \in \{0, 1, \ldots, T-1\} \text{ satisfying } |i - j| \leq L,$$

each $\left(\{\Omega_t\}_{t=0}^{T-1}\right)$-program $\{x_t\}_{t=0}^T$ such that

$$\sum_{t=0}^{T-1} \alpha_t u_t(x_t, x_{t+1}) = \sigma\left(\{\alpha_t u_t\}_{t=0}^{T-1}, \{\Omega_t\}_{t=0}^{T-1}, 0, T, x_0\right)$$

and each integer $S \in [0, T-L]$ satisfying

$$x_S \in \bar{Y}\left(\{\Omega_t\}_{t=S}^{S+l-1}, S, S+l\right) \tag{3.31}$$

the inequality

$$\min\{\rho(x_t, \bar{x}) : \ t = S+1, \ldots, S+L\} \leq \gamma \tag{3.32}$$

holds.

(ii) For each

$$\Omega_t \in \mathcal{E}(\tilde{\gamma}), \ t = 0, 1, \ldots,$$

each $u_t \in \mathcal{M}$, $t = 0, 1, \ldots$, satisfying

$$\|u_t - v\| \leq \tilde{\gamma}, \ t = 0, 1, \ldots,$$

each sequence $\{\alpha_t\}_{t=0}^\infty \subset (0, 1]$ which satisfies

$$\alpha_i \alpha_j^{-1} \geq \tilde{\lambda} \text{ for each pair of integers } i, j \geq 0 \text{ satisfying } |i - j| \leq L,$$

each $\left(\{\alpha_t u_t\}_{t=0}^\infty, \{\Omega_t\}_{t=0}^\infty\right)$-overtaking optimal program $\{x_t\}_{t=0}^\infty$ and each integer $S \geq 0$ satisfying (3.31) inequality (3.32) holds.

Choose $\lambda \in (0, 1)$ such that

$$\lambda < \tilde{\lambda},$$
$$2|1 - \lambda|\lambda^{-1}(\|v\| + 1)L < \gamma/2. \tag{3.33}$$

By property (P18), there exists $\delta \in (0, \tilde{\gamma})$ such that the following property holds:
(P19) For each natural number $S \leq 2L+4$, each

$$\Omega_t \in \mathcal{E}(\delta), \ t = 0, \ldots, S-1,$$

each $u_t \in \mathcal{M}$, $t = 0, \ldots, S-1$ satisfying

$$\|u_t - v\| \leq \delta, \ t = 0, \ldots, S-1$$

and each $\left(\{\Omega_t\}_{t=0}^{S-1}\right)$-program $\{y_t\}_{t=0}^{S}$ satisfying

$$\rho(y_0, \bar{x}), \ \rho(y_S, \bar{x}) \le \gamma$$

and

$$\sum_{t=0}^{S-1} u_t(y_t, y_{t+1}) \ge \sigma\left(\{u_t\}_{t=0}^{S-1}, \{\Omega_t\}_{t=0}^{S-1}, 0, S, y_0, y_S\right) - \gamma$$

the inequality $\rho(y_t, \bar{x}) \le \epsilon$ holds for all $t = 0, \dots, S$.

In the case of Theorem 3.2 assume that $T > 2L$ is an integer. In the case of Theorem 3.3 put $T = \infty$.

Assume that for all integers t satisfying $0 \le t < T$,

$$\Omega_t \in \mathcal{E}(\delta), \tag{3.34}$$

$u_t \in \mathcal{M}_0$ satisfies

$$\|u_t - v\| \le \delta, \tag{3.35}$$

$$\alpha_t \in (0, 1] \tag{3.36}$$

satisfy

$$\alpha_i \alpha_j^{-1} \ge \lambda \text{ for each pair of integers } i, j$$
$$\text{satisfying } 0 \le i, j < T, \ |i - j| \le L \tag{3.37}$$

and that an $\left(\{\Omega_t\}_{t=0}^{T-1}\right)$-program $\{x_t\}_{t=0}^{T}$ satisfies

$$x_0 \in \bar{Y}\left(\{\Omega_t\}_{t=0}^{l-1}, 0, l\right). \tag{3.38}$$

Assume that in the case of Theorem 3.2

$$\sigma\left(\{\alpha_t u_t\}_{t=0}^{T-1}, \{\Omega_t\}_{t=0}^{T-1}, 0, T, x_0\right) = \sum_{t=0}^{T-1} \alpha_t u_t\left(x_t, x_{t+1}\right). \tag{3.39}$$

In the case of Theorem 3.3 assume that

$$\{x_t\}_{t=0}^{\infty} \text{ is an } \left(\{\alpha_t u_t\}_{t=0}^{\infty}, \{\Omega_t\}_{t=0}^{\infty}\right) - \text{overtaking optimal program.} \tag{3.40}$$

In the case of Theorem 3.3 using (3.34)–(3.38), (3.40), and applying by induction property (ii) we obtain a sequence of integers S_i, $i = 0, 1, \dots$, such that

$$0 \le S_0 \le L \tag{3.41}$$

and for each integer $i \ge 0$,

$$1 \le S_{i+1} - S_i \le L, \tag{3.42}$$

$$\rho\left(x_{S_i}, \bar{x}\right) \le \gamma. \tag{3.43}$$

In the case of Theorem 3.2 using (3.34)–(3.39) and applying by induction property
(i) we obtain a sequence of integers S_i, $i = 0, 1, \ldots, q$ such that

$$0 \le S_0 \le L, \; T - L < S_q \le T, \tag{3.44}$$

$$1 \le S_{i+1} - S_i \le L, \; i = 0, \ldots, q - 1, \tag{3.45}$$

$$\rho\left(x_{S_i}, \bar{x}\right) \le \gamma, \; i = 0, \ldots, q. \tag{3.46}$$

In the case of Theorem 3.3 set $q = \infty$.
If $\rho(x_0, \bar{x}) \le \gamma$, we may assume that $S_0 = 0$.
Let an integer $i \ge 0$ satisfy $i + 1 \le q$. By (3.40) and (3.39),

$$\sum_{t=S_i}^{S_{i+1}-1} \alpha_t u_t \left(x_t, x_{t+1}\right) = \sigma\left(\{\alpha_t u_t\}_{t=S_i}^{S_{i+1}-1}, \{\Omega_t\}_{t=S_i}^{S_{i+1}-1}, S_i, S_{i+1}, x_{S_i}, x_{S_{i+1}}\right). \tag{3.47}$$

In view of (3.35)–(3.37) and (3.45), for each $(\{\Omega_t\}_{t=S_i}^{S_{i+1}-1})$-program $\{y_t\}_{t=S_i}^{S_{i+1}}$,

$$\left| \sum_{t=S_i}^{S_{i+1}-1} u_t(y_t, y_{t+1}) - \sum_{t=S_i}^{S_{i+1}-1} \alpha_{S_i}^{-1} \alpha_t u_t(y_t, y_{t+1}) \right|$$

$$\le (\|v\| + 1) \sum_{t=S_i}^{S_{i+1}-1} |1 - \alpha_{S_i}^{-1} \alpha_t| \le (\|v\| + 1) L |\lambda - 1| \lambda^{-1}. \tag{3.48}$$

It follows from (3.33), (3.47), and (3.48) that

$$\left| \sum_{t=S_i}^{S_{i+1}-1} u_t \left(x_t, x_{t+1}\right) - \sigma\left(\{u_t\}_{t=S_i}^{S_{i+1}-1}, \{\Omega_t\}_{t=S_i}^{S_{i+1}-1}, S_i, S_{i+1}, x_{S_i}, x_{S_{i+1}}\right) \right|$$

$$\le 2(\|v\| + 1) L |\lambda - 1| \lambda^{-1} < \gamma. \tag{3.49}$$

By (3.34), (3.35), (3.44), (3.46), (3.49), and property (P19),

$$\rho(x_t, \bar{x}) \le \epsilon \text{ for all integers } t \in [S_i, S_{i+1}]. \tag{3.50}$$

Since (3.50) holds for any nonnegative integer i satisfying $i + 1 < q$, Theorems 3.2
and 3.3 are proved.

3.4   Proof of Theorem 3.4

We recall that $\delta \in (0, \epsilon)$, $\lambda \in (0, 1)$ and the integer $L > l$ be as guaranteed by Theorem 3.2,

$$u_t \in \mathcal{M}_0 \text{ and } \|u_t - v\| \leq \delta, \ t = 0, 1, \ldots, \tag{3.51}$$

$$\Omega_t \in \mathcal{E}(\delta), \ t = 0, 1, \ldots \tag{3.52}$$

and that $\{\alpha_t\}_{t=0}^{\infty} \subset (0, 1]$ satisfy the relations

$$\lim_{t \to \infty} \alpha_t = 0, \tag{3.53}$$

$$\alpha_i \alpha_j^{-1} \geq \lambda \text{ for all nonnegative integers } i, j \text{ satisfying } |i - j| \leq L. \tag{3.54}$$

In the proof we use the following auxiliary result.

Lemma 3.7 *Let $\gamma > 0$. Then there is a natural number n_0 such that for each pair of integers $T > S \geq n_0$ and each $\{\Omega_t\}_{t=0}^{T-1}$-program $\{x_t\}_{t=0}^{T}$ satisfying*

$$x_0 \in \bar{Y}\left(\{\Omega_t\}_{t=0}^{l-1}, 0, l\right), \tag{3.55}$$

$$\sum_{t=0}^{T-1} \alpha_t u_t(x_t, x_{t+1}) = \sigma\left(\{\alpha_t u_t\}_{t=0}^{T-1}, \{\Omega_t\}_{t=0}^{T-1}, 0, T, x_0\right) \tag{3.56}$$

the following inequality holds:

$$\sum_{t=0}^{S-1} \alpha_t u_t(x_t, x_{t+1}) \geq \sigma\left(\{\alpha_t u_t\}_{t=0}^{S-1}, \{\Omega_t\}_{t=0}^{S-1}, 0, S, x_0\right) - \gamma. \tag{3.57}$$

Proof Since $\lim_{t \to \infty} \alpha_t = 0$ there exists a natural number

$$n_0 > 4L + 4$$

such that for all integers $t > n_0 - L - 4$,

$$\alpha_t \leq \gamma(8L + 8)^{-1}(\|v\| + 1)^{-1}. \tag{3.58}$$

Assume that integers $T > S \geq n_0$ and that an $\{\Omega_t\}_{t=0}^{T-1}$-program $\{x_t\}_{t=0}^{T}$ satisfies (3.55) and (3.56). By (3.51), (3.52), (3.55), and Proposition 3.1, there is an $\{\Omega_t\}_{t=0}^{S-1}$-program $\{\tilde{x}_t\}_{t=0}^{S}$ such that

$$\tilde{x}_0 = x_0, \tag{3.59}$$

$$\sum_{t=0}^{S-1} \alpha_t u_t(\tilde{x}_t, \tilde{x}_{t+1}) = \sigma\left(\{\alpha_t u_t\}_{t=0}^{S-1}, \{\Omega_t\}_{t=0}^{S-1}, 0, S, x_0\right). \tag{3.60}$$

By the choice of δ and L, Theorem 3.2, (3.51), (3.52), (3.54), (3.55), (3.56), (3.59), and (3.60),

$$\rho(x_t, \bar{x}) \leq \bar{\lambda}/4, \ t = L, \ldots, T - L, \tag{3.61}$$

$$\rho(\tilde{x}_t, \bar{x}) \leq \bar{\lambda}/4, \ t = L, \ldots, S - L. \tag{3.62}$$

By (2.4), (3.61), and (3.62) there is an $\{\Omega_t\}_{t=0}^{T-1}$-program $\{y_t\}_{t=0}^{T}$ such that

$$y_t = \tilde{x}_t, \ t = 0, \ldots, S - L, \ y_t = x_t, \ t = S - L + 1, \ldots, T. \tag{3.63}$$

In view of (3.56), (3.63), (3.60), (3.51), and (3.58),

$$0 \leq \sum_{t=0}^{T-1} \alpha_t u_t(x_t, x_{t+1}) - \sum_{t=0}^{T-1} \alpha_t u_t(y_t, y_{t+1})$$

$$= \sum_{t=0}^{S-L} \alpha_t u_t(x_t, x_{t+1}) - \sum_{t=0}^{S-L} \alpha_t u_t(y_t, y_{t+1})$$

$$\leq \sum_{t=0}^{S-L-1} \alpha_t u_t(x_t, x_{t+1}) - \sum_{t=0}^{S-L-1} \alpha_t u_t(\tilde{x}_t, \tilde{x}_{t+1}) + 2\alpha_{S-L}(\|v\| + 1)$$

$$\leq \sum_{t=0}^{S-1} \alpha_t u_t(x_t, x_{t+1}) + (\|v\| + 1) \sum_{t=S-L}^{S-1} \alpha_t$$

$$- \sigma\left(\{\alpha_t u_t\}_{t=0}^{S-1}, \{\Omega_t\}_{t=0}^{S-1}, 0, S, x_0\right) + (\|v\| + 1) \sum_{t=S-L}^{S-1} \alpha_t + 2\alpha_{S-L}(\|v\| + 1)$$

$$\leq \sum_{t=0}^{S-1} \alpha_t u_t(x_t, x_{t+1}) - \sigma\left(\{\alpha_t u_t\}_{t=0}^{S-1}, \{\Omega_t\}_{t=0}^{S-1}, 0, S, x_0\right) + \gamma.$$

Lemma 3.7 is proved.

Completion of the proof of Theorem 3.4

Let

$$z \in \bar{Y}\left(\{\Omega_t\}_{t=0}^{l-1}, 0, l\right). \tag{3.64}$$

By (2.4), (3.51), (3.52), (3.64), and Proposition 3.1, for each integer $T \geq 1$ there is an $\{\Omega_t\}_{t=0}^{T-1}$-program $\{x_t^{(z,T)}\}_{t=0}^{T}$ such that

$$x_0^{(z,T)} = z, \tag{3.65}$$

$$\sum_{t=0}^{T-1} \alpha_t u_t\left(x_t^{(z,T)}, x_{t+1}^{(z,T)}\right) = \sigma\left(\{\alpha_t u_t\}_{t=0}^{T-1}, \{\Omega_t\}_{t=0}^{T-1}, 0, T, z\right). \tag{3.66}$$

Clearly there exists a strictly increasing sequence of natural numbers $\{T_j\}_{j=1}^{\infty}$ such that for any integer $t \geq 0$ there exists

$$x_t^{(z)} = \lim_{j \to \infty} x_t^{(z,T_j)}. \tag{3.67}$$

Clearly, $\{x_t^{(z)}\}_{t=0}^{\infty}$ is an $\{\Omega_t\}_{t=0}^{\infty}$- program and

$$x_0^{(z)} = z, \tag{3.68}$$

Let $\gamma > 0$. By Lemma 3.7 there is a natural number n_0 such that the following property holds:

(P20) For each pair of integers $T > S \geq n_0$ and each $\{\Omega_t\}_{t=0}^{T-1}$-program $\{x_t\}_{t=0}^{T}$ satisfying (3.55) and (3.56), Eq. (3.57) holds.

Let $S \geq n_0$ be an integer. By (P20), (3.64), (3.65), and (3.66), for each natural number j satisfying $T_j > S$,

$$\sum_{t=0}^{S-1} \alpha_t u_t \left(x_t^{(z,T_j)}, x_{t+1}^{(z,T_j)} \right) \geq \sigma \left(\{\alpha_t u_t\}_{t=0}^{S-1}, \{\Omega_t\}_{t=0}^{S-1}, 0, S, z \right) - \gamma.$$

Together with (3.57) this implies that

$$\sum_{t=0}^{S-1} \alpha_t u_t \left(x_t^{(z)}, x_{t+1}^{(z)} \right) \geq \sigma \left(\{\alpha_t u_t\}_{t=0}^{S-1}, \{\Omega_t\}_{t=0}^{S-1}, 0, S, z \right) - \gamma$$

for all integers $S \geq n_0$. Theorem 3.4 is proved.

Chapter 4
Optimal Control Problems with Nonsingleton Turnpikes

In this chapter we study stability of the turnpike phenomenon for approximate solutions for a general class of discrete-time optimal control problems with nonsingleton turnpikes and with a compact metric space of states. This class of optimal control problems is identified with a complete metric space of objective functions. We show that the turnpike phenomenon is stable under perturbations of an objective function if the corresponding infinite horizon optimal control problem possesses an asymptotic turnpike property.

4.1 Discrete-Time Optimal Control Systems

Let $(\mathcal{K}, d)$ be a compact metric space equipped with a metric d and with the topology induced by this metric, and let $\mathcal{M}$ be a nonempty closed subset of $\mathcal{K} \times \mathcal{K}$ equipped with the product topology. Set

$$A = \{x \in \mathcal{K} : \{x\} \times \mathcal{K} \cap \mathcal{M} \neq \emptyset\} \tag{4.1}$$

and

$$a(x) = \{y \in \mathcal{K} : (x, y) \in \mathcal{M}\} \text{ for all } x \in A. \tag{4.2}$$

Denote by $\mathbf{Z}$ the set of all integers. For each pair of integers $q > p$ set

$$\mathbf{Z}_p = \{i \in \mathbf{Z} : i \geq p\} \text{ and } \mathbf{Z}_p^q = [p, q] \cap \mathbf{Z}. \tag{4.3}$$

A sequence $x_i \in \mathcal{K}$, $i \in I$, where I is either $\mathbf{Z}$ or $\mathbf{Z}_p$ or $\mathbf{Z}_p^q$ (with $p, q \in \mathbf{Z}$ satisfying $p < q$) is called a program if $(x_i, x_{i+1}) \in \mathcal{M}$ for each integer $i \in I$ such that $i + 1 \in I$.

Denote by $C(\mathcal{M})$ the space of all continuous functions $v : \mathcal{M} \to R^1$ and by $B(\mathcal{M})$ the space of all bounded functions $v : \mathcal{M} \to R^1$ with the topology of the uniform convergence ($\|v\| = \sup\{|v(x, y)| : (x, y) \in \mathcal{M}\}$).

A. J. Zaslavski, *Stability of the Turnpike Phenomenon in Discrete-Time Optimal Control Problems*, SpringerBriefs in Optimization, DOI 10.1007/978-3-319-08034-5_4, © The Author 2014

In this chapter we consider the problem

$$\sum_{i=0}^{T-1} v(x_i, x_{i+1}) \rightarrow \min \tag{P}$$

$$\text{s.t. } \{(x_i, x_{i+1})\}_{i=0}^{T-1} \subset \mathcal{M},$$

where T is a natural number and $v \in B(\mathcal{M})$ is a lower semicontinuous function. This discrete-time optimal control system describes a general model of economic dynamics, where the set $\mathcal{K}$ is the space of states, $-v$ is a utility function and $-v(x_t, x_{t+1})$ evaluates consumption at moment t.

For each $f \in B(\mathcal{M})$, each $y, z \in \mathcal{K}$ and each integer $q \geq 1$ we set

$$U^f(q, y, z) = \inf \left\{ \sum_{i=0}^{q-1} f(x_i, x_{i+1}) : \right.$$

$$\{x_i\}_{i=0}^{q} \text{ is a program such that } x_0 = y, \ x_q = z\}. \tag{4.4}$$

Let $T_1 \geq 0$, $T_2 > T_1$ be integers, $y, z \in \mathcal{K}$, and let $f_i \in B(\mathcal{M}), i = T_1, \ldots, T_2 - 1$. Set

$$U(\{f_i\}_{i=T_1}^{T_2-1}, y, z) = \inf \left\{ \sum_{i=T_1}^{T_2-1} f_i(x_i, x_{i+1}) : \right.$$

$$\{x_i\}_{i=T_1}^{T_2} \text{ is a program such that } x_{T_1} = y, \ x_{T_2} = z\}. \tag{4.5}$$

(We suppose that infimum over an empty set is ∞.)

For any subset E of a metric space the closure of E is denoted by $\mathrm{cl}(E)$ and also by $\bar{E}$.

For any sequence $\{x_i\}_{i=0}^{\infty} \subset \mathcal{K}$ denote by $\Omega\left(\{x_i\}_{i=0}^{\infty}\right)$ the set of all points $(z_1, z_2) \in \mathcal{K} \times \mathcal{K}$ such that some subsequence $\{(x_{i_k}, x_{i_k+1})\}_{k=1}^{\infty}$ converges to (z_1, z_2) and denote by $\omega(\{x_i\}_{i=0}^{\infty})$ the set of all points $z \in \mathcal{K}$ such that some subsequence $\{x_{i_k}\}_{k=1}^{\infty}$ converges to z.

Define a metric d_1 on $\mathcal{K} \times \mathcal{K}$ by

$$d_1((x_1, x_2), (y_1, y_2)) = d(x_1, y_1) + d(x_2, y_2), \ x_1, x_2, y_1, y_2 \in \mathcal{K}.$$

Put $d(x, B) = \inf\{d(x, y) : y \in B\}$ for $x \in \mathcal{K}, B \subset \mathcal{K}$ and

$$d_1((x_1, x_2), E) = \inf\{d_1((x_1, x_2), (y_1, y_2)) : (y_1, y_2) \in E\}$$

for $(x_1, x_2) \in \mathcal{K} \times \mathcal{K}$ and $E \subset \mathcal{K} \times \mathcal{K}$.

We denote by $\mathrm{dist}(B_1, B_2)$ the Hausdorff metric for two sets $B_1, B_2 \subset \mathcal{K}$ (respectively $B_1, B_2 \subset \mathcal{K} \times \mathcal{K}$) and denote by $\mathrm{Card}(B)$ the cardinality of a set B.

4.2 The Turnpike Property

Denote by $\mathfrak{M}_{reg}$ the set of all lower semicontinuous functions $f \in B(\mathcal{M})$ which satisfy the following assumption.

(A) There exist a program $\{z_j^f\}_{j=0}^{\infty} \subset \mathcal{K}$, constants $c(f) > 0$, $\gamma(f) > 0$, and $\mu(f) \in R^1$ and an open set $V_f \subset \mathcal{M}$ in the relative topology such that:

    (i) $|\sum_{j=0}^{N-1} [f(z_j^f, z_{j+1}^f) - \mu(f)]| \leq c(f)$ for all integers $N \geq 1$.

    (ii) For each integer $N \geq 1$ and each program $\{z_j\}_{j=1}^{N} \subset \mathcal{K}$

$$\sum_{j=0}^{N-1} [f(z_j, z_{j+1}) - \mu(f)] \geq -c(f).$$

    (iii) $[\omega(\{z_j^f\}_{j=0}^{\infty})]^2 \subset V_f$ and the restriction $f|V_f : V_f \to R^1$ is a continuous function.

    (iv) For each integer $j \geq 0$ and each $(x, y) \in \mathcal{M}$ satisfying

$$d_1\left((x, y), \left(z_j^f, z_{j+1}^f\right)\right) \leq \gamma(f)$$

the inclusions (x, z_{j+1}^f), $(z_j^f, y) \in \mathcal{M}$ hold.

    (v) For each integer $j \geq 0$ and each $(x, y) \in \mathcal{M}$ satisfying $d(x, z_{j+1}^f) \leq \gamma(f)$ the inclusion $(z_j^f, x) \in \mathcal{M}$ holds.

    (vi) For each $x, y, z \in \omega(\{z_j^f\}_{j=0}^{\infty})$ which satisfy $(x, y) \in \mathcal{M}$ and $d(x, z) \leq \gamma(f)$ the inclusion $(z, y) \in \mathcal{M}$ holds.

Let $f \in B(\mathcal{M})$. Clearly, (A)(iii) holds if the function f is continuous.

Assumptions (A)(iv)–(A)(vi) hold if there is $\gamma_0 > 0$ such that all the closed balls in $\mathcal{K} \times \mathcal{K}$ with radius γ_0 and with centers belonging to $\mathcal{M} \cap [cl(\{z_j^f\}_{j=0}^{\infty})]^2$ are contained in $\mathcal{M}$.

Assumptions (A)(i) and (A)(ii) imply that for any natural number T

$$\sum_{j=0}^{T-1} f\left(z_j^f, z_{j+1}^f\right) \leq \inf \{U^f(T, y, z): \ y, z \in \mathcal{K}\} + 2c(f).$$

It means that the program $\{z_j^f\}_{j=0}^{\infty}$ is an approximate (up to $2c(f)$) solution of problem (P) with any natural number T. It should be mentioned that a program which possesses this property usually exists for optimal control problems with the turnpike property [45].

Let $\mathfrak{A}$ be either $\mathfrak{M}_{reg}$ or $\mathfrak{M}_{reg} \cap C(\mathcal{M})$.

Denote by $\bar{\mathfrak{A}}$ the closure of $\mathfrak{A}$ in $B(\mathcal{M})$ and consider the topological space $\bar{\mathfrak{A}}$ with the relative topology.

In [53] we proved the existence of a set $\mathcal{F} \subset \bar{\mathfrak{A}}$ which is a countable intersection of open everywhere dense sets in $\bar{\mathfrak{A}}$ and for which the following theorem is true.

Theorem 4.1 *Let $f \in \mathcal{F}$. Then there exists a nonempty closed set $H(f) \subset \mathcal{M}$ such that for each $S > 0$ and each $\epsilon > 0$ there exist a neighborhood $\mathcal{U}$ of f in $B(\mathcal{M})$ and integers $l, L, Q \geq 1$ such that the following assertion holds:*

For each $g \in \mathcal{U}$, each integer $T \geq L + lQ$. and each program $\{x_i\}_{i=0}^{T} \subset \mathcal{K}$ which satisfies

$$\sum_{i=0}^{T-1} g(x_i, x_{i+1}) \leq \inf\{U^g(T, y, z) : y, z \in \mathcal{K}\} + S,$$

there exist sequences of integers $\{b_i\}_{i=1}^{q}$, $\{c_i\}_{i=1}^{q} \subset [0, T]$ such that

$$q \leq Q, \; 0 \leq c_i - b_i \leq l, \; i = 1, \ldots q,$$
$$\mathrm{dist}(H(f), \{(x_i, x_{i+1}) : i = p, \ldots, p + L - 1\}) \leq \epsilon$$

for each integer $p \in [0, T - L] \setminus \cup_{i=1}^{q}[b_i, c_i]$.

This result shows that any $f \in \mathcal{F}$ possesses the turnpike property with the turnpike $H(f)$ which is not necessarily a singleton and that this turnpike property is stable under perturbations of the objective function.

Example 4.2 Suppose that $\mathcal{M} = \mathcal{K} \times \mathcal{K}$. It follows from the results of Leizarowitz [22] that $C(\mathcal{K} \times \mathcal{K}) \subset \mathfrak{M}_{reg}$. Then there exists a set $\mathcal{F} \subset C(\mathcal{K} \times \mathcal{K})$ which is a countable intersection of open everywhere dense subsets of $C(\mathcal{K} \times \mathcal{K})$ and for which Theorem 4.1 is valid.

Let $x_0, y_0 \in \mathcal{K}$ and $x_0 \neq y_0$. Define a function $g \in C(\mathcal{K} \times \mathcal{K})$ by

$$g(x, y) = d(x, x_0)d(x, y_0), \; x, y \in K.$$

Clearly, the function g does not have the turnpike property. It means that Theorem 4.1 cannot be improved in principle.

Example 4.3 Let $\mathcal{K}$ be a compact convex subset of R^n and let $\mathcal{M} \subset \mathcal{K} \times \mathcal{K}$ be a convex compact subset of $R^n \times R^n$ with an nonempty interior denoted by $\mathrm{int}(\mathcal{M})$. We assume that $d(x, y) = \|x - y\|$, $x, y \in \mathcal{K}$ where $\| \cdot \|$ is the Euclidean norm in R^n induced by an inner product $\langle \cdot, \cdot \rangle$.

Denote by $\mathfrak{M}_{conv}$ the set of all convex lower semicontinuous functions $f \in B(\mathcal{M})$ for which there exists $(z_f, z_f) \in \mathrm{int}(\mathcal{M})$ such that

$$f(z, z) \geq f(z_f, z_f) \text{ for all } (z, z) \in \mathcal{M}. \tag{4.6}$$

Let $f \in \mathfrak{M}_{conv}$. It is a well-known fact of convex analysis [37] that there exists $\eta \in R^n$ such that

$$f(x, y) \geq f(z_f, z_f) + \langle \eta, x - y \rangle \text{ for all } (x, y) \in \mathcal{M}. \tag{4.7}$$

It follows from (4.6) and (4.7) that $\mathfrak{M}_{conv} \subset \mathfrak{M}_{reg}$.

Let $\mathfrak{A}$ be either $\mathfrak{M}_{reg}$ or $\mathfrak{M}_{reg} \cap C(\mathcal{M})$, and let $\mathcal{F} \subset \bar{\mathfrak{A}}$ be a countable intersection of open everywhere dense subsets of $\bar{\mathfrak{A}}$ for which Theorem 4.1 is true. It follows from the construction of the set $\mathcal{F}$ (see [53]) that the set $\mathcal{F} \cap \mathrm{cl}(\mathfrak{M}_{conv} \cap \mathfrak{A})$ is a countable intersection of open everywhere dense sets in $\mathrm{cl}(\mathfrak{M}_{conv} \cap \mathfrak{A})$. Therefore, a generic function $f \in \mathrm{cl}(\mathfrak{M}_{conv} \cap \mathfrak{A})$ has the turnpike property.

Let $f \in \mathfrak{M}_{reg}$. There exist a program $\{z_j^f\}_{j=0}^{\infty} \subset \mathcal{K}$, constants $c(f) > 0$, $\gamma(f) > 0$, and $\mu(f) \in R^1$, and an open set $V_f \subset \mathcal{M}$ in the relative topology such that assumption (**A**) holds.

A program $\{y_j\}_{j=0}^{\infty} \subset \mathcal{K}$ is called (f)-good [22, 45] if

$$\sup \left\{ \left| \sum_{j=0}^{N-1} [f(y_j, y_{j+1}) - \mu(f)] \right| : N = 1, 2, \ldots \right\} < \infty.$$

We can easily deduce the following result.

Proposition 4.4 *Let* $\{y_j\}_{j=0}^{\infty} \subset \mathcal{K}$ *be a program. Then either* $\{y_j\}_{j=0}^{\infty}$ *is an* (f)-*good program or*

$$\sum_{j=0}^{N-1} [f(y_j, y_{j+1}) - \mu(f)] \to \infty \text{ as } N \to \infty.$$

Moreover, if $\{y_j\}_{j=0}^{\infty}$ *is an* (f)-*good program, then there exists an integer* $N_0 \geq 1$ *such that for each pair of integers* $q > p \geq N_0$,

$$\sum_{j=p}^{q-1} [f(y_j, y_{j+1}) - \mu(f)] \leq c(f) + 1.$$

We say that f possesses the asymptotic turnpike property or, briefly, (ATP) if for each pair of (f)-good programs $\{y_j\}_{j=0}^{\infty}$, $\{y_j'\}_{j=0}^{\infty}$,

$$\Omega \left(\{y_j\}_{j=0}^{\infty} \right) = \Omega \left(\{y_j'\}_{j=0}^{\infty} \right).$$

In this case there exists a nonempty compact set $H(f) \subset \mathcal{M}$ depending only on f such that

$$\Omega \left(\{y_j\}_{j=0}^{\infty} \right) = H(f)$$

for any (f)-good program $\{y_j\}_{j=0}^{\infty}$.

In this chapter we prove the following result which shows that the turnpike phenomenon holds and is stable under perturbation of the objective function f if it possesses the asymptotic turnpike property.

Theorem 4.5 *Assume that* $f \in \mathfrak{M}_{reg}$ *and that there exists a nonempty closed set* $H(f) \subset \mathcal{M}$ *such that*

$$\Omega \left(\{y_j\}_{j=0}^{\infty} \right) = H(f)$$

for any (f)-*good program* $\{y_j\}_{j=0}^{\infty}$.

Let $\epsilon, M > 0$. *Then there exist a neighborhood* $\mathcal{U}$ *of* f *in* $B(\mathcal{M})$ *and integers* $l, L, Q \geq 1$ *such that the following assertion holds:*

For each integer $T \geq L + lQ$, each $g_i \in \mathcal{U}$, $i = 0, \ldots, T-1$ and each program $\{x_i\}_{i=0}^T$ which satisfies

$$\sum_{i=0}^{T-1} g_i(x_i, x_{i+1}) \leq \inf \left\{ U\left(\{g_i\}_{i=0}^{T-1}, y, z\right) : y, z \in \mathcal{K} \right\} + M$$

there exist sequences of integers $\{b_i\}_{i=1}^q$, $\{c_i\}_{i=1}^q \subset [0, T]$ such that

$$q \leq Q, \ 0 \leq c_i - b_i \leq l, \ i = 1, \ldots q,$$

$$dist(H(f), \{(x_i, x_{i+1}) : \ i = p, \ldots, p + L - 1\}) \leq \epsilon$$

for each integer $p \in [0, T-L] \setminus \cup_{i=1}^q [b_i, c_i]$.

4.3 Preliminaries

Let $f \in \mathfrak{M}_{reg}$. There exist a program $\{z_j^f\}_{j=0}^\infty \subset \mathcal{K}$, constants $c(f) > 0$, $\gamma(f) > 0$ and $\mu(f) \in R^1$ and an open set $V_f \subset \mathcal{M}$ in the relative topology such that assumption (**A**) holds.

We can easily deduce the following three results.

Proposition 4.6 *Assume that $x, y \in \mathcal{K}$, $q \geq 1$ is an integer, and that $U^f(q, x, y) < +\infty$. Then there exists a program $\{z_j\}_{j=0}^q \subset K$ such that*

$$z_0 = x, \ z_q = y, \ \sum_{j=0}^{q-1} f(z_j, z_{j+1}) = U^f(q, x, y).$$

Proposition 4.7 *For each integer $q \geq 1$ the function $(x, y) \to U^f(q, x, y)$, $x, y \in \mathcal{K}$ is lower semicontinuous.*

A program x_j, $j \in I$, where I is either $\mathbf{Z}$ or $\mathbf{Z}_p$ or $\mathbf{Z}_p^q$ (with $p, q \in \mathbf{Z}$ satisfying $p < q$) is (f)-minimal [6, 24, 41, 45] if for each $n, m \in I$ satisfying $m < n$

$$\sum_{j=m}^{n-1} f(x_j, x_{j+1}) = U^f(n - m, x_m, x_n).$$

Proposition 4.8 *1. For each $(x, y) \in \mathcal{M}$ and each $(\tilde{x}, \tilde{y}) \in \Omega(\{z_i^f\}_{i=0}^\infty)$ satisfying $d_1((x, y), (\tilde{x}, \tilde{y})) < \gamma(f)$ the inclusions $(x, \tilde{y}), (\tilde{x}, y) \in \mathcal{M}$ hold.*

2. For each $(x, y) \in \mathcal{M}$ and each $(\tilde{x}, \tilde{y}) \in \Omega(\{z_i^f\}_{i=0}^\infty)$ satisfying $d(\tilde{y}, x) < \gamma(f)$ the inclusion $(\tilde{x}, x) \in \mathcal{M}$ holds.

Proposition 4.9 *Let $\{z_i\}_{i=0}^\infty \subset K$ be an (f)-good program and let*

$$(y_0, y_1) \in \Omega\left(\{z_i\}_{i=0}^\infty\right).$$

Then there exists a program $\{y_i\}_{i=-\infty}^{\infty} \subset \mathcal{K}$ *such that*

$$(y_j, y_{j+1}) \in \Omega\left(\{z_i\}_{i=0}^{\infty}\right) \text{ for all } j \in \mathbf{Z}, \tag{4.8}$$

$$\sum_{j=p}^{q-1} \left[f\left(y_j, y_{j+1}\right) - \mu(f)\right] \le c(f) + 1 \text{ for all integers } q > p. \tag{4.9}$$

Moreover, if $\Omega(\{z_i\}_{i=0}^{\infty}) \subset \Omega(\{z_i^f\}_{i=0}^{\infty})$, *then* $\{y_i\}_{i=-\infty}^{\infty}$ *is an* (f)-*minimal program.*

Proof There exists a subsequence

$$(z_{i_k}, z_{i_k+1}) \to (y_0, y_1) \text{ as } k \to \infty. \tag{4.10}$$

By Proposition 4.4 we may assume without loss of generality that

$$\sum_{j=p}^{q-1} \left[f\left(z_j, z_{j+1}\right) - \mu(f)\right] \le c(f) + 1 \text{ for all nonnegative integers } q > p. \tag{4.11}$$

For each integer $k \ge 1$ we set

$$y_j^k = z_{j+i_k} \text{ for all integers } j \ge -i_k. \tag{4.12}$$

There exist a subsequence of programs $\{y_i^{k_j} : i \ge -i_{k_j}\}$, $j = 1, 2, \ldots$ and a sequence $\{y_i\}_{i=-\infty}^{\infty}$ such that

$$y_i^{k_j} \to y_i \text{ as } j \to \infty \text{ for each integer } i. \tag{4.13}$$

It is easy to see that the sequence $\{y_i\}_{i=-\infty}^{\infty}$ is a program. Equations (4.8) and (4.9) follow from (4.11), (4.12), (4.13), and the lower semicontinuity of f.

Assume that

$$\Omega\left(\{z_i\}_{i=0}^{\infty}\right) \subset \Omega\left(\{z_i^f\}_{i=0}^{\infty}\right). \tag{4.14}$$

We will show that $\{y_i\}_{i=-\infty}^{\infty}$ is an (f)-minimal program.

Let us assume the contrary. Then there exist $r > 0$ and integers $p < q$ such that

$$\sum_{i=p}^{q-1} f(y_i, y_{i+1}) > U^f(q - p, y_p, y_q) + r. \tag{4.15}$$

By Proposition 4.6 there exists a program $\{x_i\}_{i=p}^{q} \subset \mathcal{K}$ such that

$$x_i = y_i, \ i = p, q, \ \sum_{i=p}^{q-1} f(x_i, x_{i+1}) = U^f(q - p, y_p, y_q). \tag{4.16}$$

By assumption (**A**)(iii), (4.14), and (4.8) there exists

$$\epsilon \in (0, 8^{-1}\gamma(f)) \tag{4.17}$$

such that for each $i \in \{p-1, \ldots, q\}$ and each $(x, y) \in \mathcal{M}$ satisfying

$$d_1((x, y), (y_i, y_{i+1})) \leq 4\epsilon \tag{4.18}$$

the inequality $|f(x, y) - f(y_i, y_{i+1})| \leq 8^{-1} r(q - p + 2)^{-1}$ holds.

Since $\{z_i\}_{i=0}^{\infty}$ is an (f)-good program there exists an integer $N_0 \geq 1$ such that

$$\sum_{i=m}^{n-1} f(z_i, z_{i+1}) \leq U(n-m, z_m, z_n) + 32^{-1} r \text{ for all integers } n > m \geq N_0. \tag{4.19}$$

By (4.12) and (4.13) there exists an integer $k \geq 1$ such that

$$i_k - |p| - |q| \geq 2N_0 + 2, \; d(z_{j+i_k}, y_j) \leq 8^{-1}\epsilon, \; j = p - 1, \ldots, q + 1. \tag{4.20}$$

It follows from (4.20), (4.8), (4.14), (4.17), and Proposition 4.8 that

$$(z_{p-1+i_k}, y_p), \; (y_q, z_{q+1+i_k}) \in \mathcal{M}. \tag{4.21}$$

We define

$$v_{p-1+i_k} = z_{p-1+i_k} \; v_j = x_{j-i_k}, \; j = p + i_k, \ldots, q + i_k, v_{q+1+i_k} = z_{q+1+i_k}. \tag{4.22}$$

Equations (4.21), (4.16), and (4.22) imply that $\{v_j : \; j = p - 1 + i_k, \ldots, q + 1 + i_k\}$ is a program. We will estimate

$$\sum_{j=p-1+i_k}^{q+i_k} [f(z_j, z_{j+1}) - f(v_j, v_{j+1})].$$

By (4.22), (4.19), and (4.20),

$$\sum_{j=p-1+i_k}^{q+i_k} [f(z_j, z_{j+1}) - f(v_j, v_{j+1})] \leq 32^{-1} r. \tag{4.23}$$

On the other hand it follows from (4.22), (4.16), and (4.15) that

$$\sum_{j=p-1+i_k}^{q+i_k} [f(z_j, z_{j+1}) - f(v_j, v_{j+1})] = f(z_{p-1+i_k}, z_{p+i_k}) - f(z_{p+i_k-1}, y_p)$$

$$+ \sum_{j=p+i_k}^{q+i_k-1} f(z_j, z_{j+1}) - \sum_{j=p}^{q-1} f(y_j, y_{j+1}) + \sum_{j=p}^{q-1} f(y_j, y_{j+1}) - \sum_{j=p}^{q-1} f(x_j, x_{j+1})$$

$$+ f(z_{q+i_k}, z_{q+i_k+1}) - f(y_q, z_{q+i_k+1}) \geq r + f(z_{p-1+i_k}, z_{p+i_k}) - f(z_{p+i_k-1}, y_p)$$

$$+ f(z_{q+i_k}, z_{q+i_k+1}) - f(y_q, z_{q+i_k+1}) + \sum_{j=p}^{q-1} [f(z_{j+i_k}, z_{j+i_k+1}) - f(y_j, y_{j+1})].$$

By the equation above, (4.20), (4.21), and the definition of ϵ (see (4.17), (4.18)),

$$\sum_{j=p-1+i_k}^{q+i_k} [f(z_j, z_{j+1}) - f(v_j, v_{j+1})] \geq r - 2(q - p + 2)8^{-1}r(q - p - 2)^{-1} \geq 2^{-1}r.$$

This contradicts (4.23). Therefore $\{y_i\}_{-\infty}^{\infty}$ is an (f)-minimal trajectory. The proposition is proved.

Assume that f possesses (ATP) and that $H(f)$ is a nonempty closed subset of $\mathcal{M}$ such that

$$H(f) = \Omega(\{z_i\}_{i=0}^{\infty}) \text{ for any}(f) - \text{good program}\{z_i\}_{i=0}^{\infty}. \tag{4.24}$$

By Proposition 4.9 there exists an (f)-minimal program $\{x_j^f\}_{j=-\infty}^{\infty} \subset \mathcal{K}$ such that

$$(x_j^f, x_{j+1}^f) \in H(f) \text{ for each integer}j, \tag{4.25}$$

$$\sum_{j=p}^{q-1} \left[f\left(x_j^f, x_{j+1}^f\right) - \mu(f) \right] \leq c(f) + 1 \text{ for each pair of integers } q > p. \tag{4.26}$$

It follows from (ATP), assumption (A) and (4.26) that

$$\Omega \left(\left\{ x_j^f \right\}_{j=0}^{\infty} \right) = H(f) = \Omega \left(\left\{ z_j^f \right\}_{j=0}^{\infty} \right). \tag{4.27}$$

Set

$$H_0^f = \{x \in \mathcal{K} : \text{ there exists } y \in \mathcal{K} \text{ such that } (x, y) \in H(f)\}. \tag{4.28}$$

For any $x \in H_0^f$ we define

$$\pi^f(x) = \inf \left\{ \liminf_{N \to \infty} \sum_{i=0}^{N-1} [f(y_i, y_{i+1}) - \mu(f)] : \right.$$

$$\left. \text{a program } \{y_i\}_{i=0}^{\infty} \subset H_0^f \text{ and } y_0 = x \right\}. \tag{4.29}$$

Proposition 4.10 $\pi^f : H_0^f \to R^1$ *is a continuous function.*

Proof It follows from assumption (A)(ii), Proposition 4.9, and (4.25)–(4.28) that

$$\pi^f(x) \in [-c(f), c(f) + 1], \ x \in H_0^f.$$

By assumption (A)(vi) for each $x, y, z \in H_0^f$ which satisfy

$$(x, y) \in \mathcal{M}, \ d(z, x) \leq \gamma(f) \tag{4.30}$$

the inclusion $(z, y) \in \mathcal{M}$ holds. By assumption $(\mathbf{A})$(iii) and (4.27) there exists $\delta_0 \in$ $(0, 2^{-1}\gamma(f))$ such that for each $(x_1, y_1) \in H_0^f \times H_0^f$ and each $(x_2, y_2) \in \mathcal{M}$ satisfying

$$d_1((x_1, y_1), (x_2, y_2)) \le \delta_0$$

the inclusion $(x_2, y_2) \in V_f$ holds.

Set

$$Q = \left\{ (x, y) \in \mathcal{M} : d_1\left((x, y), H_0^f \times H_0^f\right) \le \delta_0 \right\}. \tag{4.31}$$

It follows from the choice of δ_0 that $Q \subset V_f$.

Let ϵ be a positive number. Since $f|V_f : V_f \to R^1$ is a continuous function (see assumption $(\mathbf{A})$(iii)) there exists $\delta \in (0, 2^{-1}\delta_0)$ such that for each $(x_1, y_1), (x_2, y_2) \in Q$ satisfying

$$d_1((x_1, y_1), (x_2, y_2)) \le \delta \tag{4.32}$$

the following inequality holds:

$$|f(x_1, y_1) - f(x_2, y_2)| \le 8^{-1}\epsilon. \tag{4.33}$$

Suppose that

$$x, y \in H_0^f, \ d(x, y) \le \delta. \tag{4.34}$$

We show that $|\pi^f(x) - \pi^f(y)| \le \epsilon$. There exists a program $\{x_j\}_{j=1}^\infty \subset H_0^f$ such that

$$x_0 = x, \ \pi^f(x) \ge \liminf_{N \to \infty} \sum_{i=0}^{N-1} [f(x_i, x_{i+1}) - \mu(f)] - 8^{-1}\epsilon. \tag{4.35}$$

It follows from (4.35), (4.34), (4.30), and the choice of δ_0, δ (see (4.31)–(4.33)) that

$$(y, x_1) \in \mathcal{M}, \ |f(y, x_1) - f(x, x_1)| \le 8^{-1}\epsilon. \tag{4.36}$$

Define

$$y_0 = y, \ y_i = x_i, \ i = 1, 2, \ldots. \tag{4.37}$$

By (4.36), (4.37), (4.34), and (4.35), $\{y_i\}_{i=0}^\infty$ is a program and

$$\pi^f(y) \le \liminf_{N \to \infty} \sum_{i=0}^{N-1} [f(y_i, y_{i+1}) - \mu(f)]$$

$$\le \liminf_{N \to \infty} \sum_{i=0}^{N-1} [f(x_i, x_{i+1}) - \mu(f)] + 8^{-1}\epsilon \le \pi^f(x) + 4^{-1}\epsilon.$$

This completes the proof of the proposition.

Proposition 4.11 *Define*

$$\theta^f(x, y) = f(x, y) - \mu(f) + \pi^f(y) - \pi^f(x) \tag{4.38}$$

for each $(x, y) \in \mathcal{M} \cap (H_0^f \times H_0^f)$. *Then* $\theta^f : \mathcal{M} \cap (H_0^f \times H_0^f) \to R^1$ *is a continuous nonnegative function such that* $\theta^f(x, y) = 0$ *for each* $(x, y) \in H(f)$.

Proof The continuity of θ^f follows from Proposition 4.10, (4.27), (4.28), and assumption (**A**)(iii). It follows from the definition of π^f, θ^f that

$$\theta^f(x, y) \geq 0$$

for each $(x, y) \in (H_0^f \times H_0^f) \cap \mathcal{M}$.

Let $(x, y) \in H(f)$. We show that $\theta^f(x, y) = 0$. By (4.25) and (4.26) for each integer $N \geq 1$,

$$c(f) + 1 \geq \sum_{i=0}^{N-1} \left[f\left(x_i^f, x_{i+1}^f\right) - \mu(f) \right] = \sum_{i=0}^{N-1} \theta^f\left(x_i^f, x_{i+1}^f\right) - \pi^f\left(x_N^f\right) + \pi^f\left(x_0^f\right),$$

$$\theta^f(x_i^f, x_{i+1}^f) \to 0 \text{ as } i \to \infty. \tag{4.39}$$

It follows from the continuity of θ^f, (4.39) and (4.27) that $\theta^f(x, y) = 0$. The proposition is proved.

Proposition 4.12 *Let* $j \in Z$. *Then*

$$\pi^f(x_j^f) = \liminf_{N \to \infty} \sum_{i=j}^{N-1} \left[f\left(x_i^f, x_{i+1}^f\right) - \mu(f) \right], \tag{4.40}$$

$$\sup\{\pi^f(y) : y \in H_0^f\} = 0.$$

Proof It follows from Propositions 4.10 and 4.11, (4.25), (4.27), and (4.28) that

$$\liminf_{N \to \infty} \sum_{i=j}^{N-1} \left[f\left(x_i^f, x_{i+1}^f\right) - \mu(f) \right] = \liminf_{N \to \infty} \left[\pi^f\left(x_j^f\right) - \pi^f\left(x_N^f\right) \right]$$

$$= \pi^f(x_j^f) - \sup\{\pi^f(y) : y \in H_0^f\}. \tag{4.41}$$

Assume that a program

$$\{y_i\}_{i=0}^\infty \subset H_0^f,$$

$$y_0 = x_j^f, \ \liminf_{N \to \infty} \sum_{i=0}^{N-1} [f(y_i, y_{i+1}) - \mu(f)] \leq \pi^f\left(x_j^f\right) + 1. \tag{4.42}$$

By Proposition 4.11 and (4.42)

$$\liminf_{N \to \infty} \sum_{i=0}^{N-1} [f(y_i, y_{i+1}) - \mu(f)] \geq \liminf_{N \to \infty} [\pi^f\left(x_j^f\right) - \pi^f(y_N)]$$

$$\geq \pi^f\left(x_j^f\right) - \sup\left\{\pi^f(y) : y \in H_0^f\right\}.$$

Since the equation above holds for each program $\{y_i\}_{i=0}^{\infty}$ satisfying (4.42) we conclude that

$$\pi^f\left(x_j^f\right) \geq \pi^f\left(x_j^f\right) - \sup\left\{\pi^f(y): \ y \in H_0^f\right\}.$$

Together with (4.41) this implies (4.40). The proposition is proved.

4.4 Auxiliary Results

Assume that $f \in \mathfrak{M}_{reg}$ possesses (ATP) and that $H(f)$ is a nonempty closed subset of $\mathcal{M}$ such that

$$H(f) = \Omega(\{z_i\}_{i=0}^{\infty}) \text{ for any } (f) - \text{ good program } \{z_i\}_{i=0}^{\infty}.$$

It follows from assumption (A) and the results of Sect. 4.3 (see (4.25)–(4.27), Proposition 4.8) that there exist an (f)-minimal program $\{x_i^f\}_{i=-\infty}^{\infty}$, constants $c(f) > 0$, $\gamma(f) > 0$, $\mu(f) \in R^1$, an open set $V_f \subset \mathcal{M}$ in the relative topology such that the following properties hold:

(a) $\sum_{j=p}^{q-1}[f(x_j^f, x_{j+1}^f) - \mu(f)] \leq c(f)$ for each pair of integers $q > p$.
(b) For each integer $N \geq 1$ and each program $\{z_j\}_{j=0}^{N} \subset \mathcal{K}$,

$$\sum_{j=0}^{N-1}\left[f\left(z_j, z_{j+1}\right) - \mu(f)\right] \geq -c(f).$$

(c) $[\omega(\{x_i^f\}_{i=0}^{\infty})]^2 \subset V_f$ and the restriction $f|_{V_f} : V_f \to R^1$ is a continuous function;
(d) For each $(x_1, y_1) \in H(f)$ and each $(x_2, y_2) \in \mathcal{M}$ satisfying

$$d_1((x_1, y_1), (x_2, y_2)) < \gamma(f)$$

the relations $(x_1, y_2), (x_2, y_1) \in \mathcal{M}$ hold.
(e) For each $(x_1, y_1) \in H(f)$ and $(x_2, y_2) \in \mathcal{M}$ satisfying $d(y_1, x_2) < \gamma(f)$ the inclusion $(x_1, x_2) \in \mathcal{M}$ holds.
(f) For each $x, y, z \in \omega(\{x_i^f\}_{i=0}^{\infty})$ which satisfy $(x, y) \in \mathcal{M}$ and $d(x, z) \leq \gamma(f)$ the inclusion $(z, y) \in \mathcal{M}$ holds.
(g) $(x_j^f, x_{j+1}^f) \in H(f)$ for each integer j, $\Omega(\{x_j^f\}_{j=0}^{\infty}) = H(f)$.

Consider the set $H_0^f \subset \mathcal{K}$ defined by (4.28) and the continuous functions $\pi^f : H_0^f \to R^1$ and $\theta^f : (H_0^f \times H_0^f) \cap \mathcal{M} \to [0, \infty)$ defined by (4.29) and (4.38).

Lemma 4.13 *Let $\epsilon > 0$. Then there exists an integer $L \geq 1$ such that for each (f)-good program $\{x_i\}_{i=0}^{\infty} \subset \mathcal{K}$,*

$$\text{dist}(H(f), \{(x_i, x_{i+1}): \ i \in [p, p+L]\}) \leq \epsilon \text{ for all sufficiently large integers } p.$$

Proof Let us assume the contrary. Then for each integer $N \geq 1$ there exists an (f)-good program $\{x_i^N\}_{i=0}^{\infty} \subset \mathcal{K}$ such that

$$\limsup_{p \to \infty} \text{dist}\left(H(f), \{(x_i^N, x_{i+1}^N) : i \in [p, p+N]\}\right) \geq \epsilon. \tag{4.43}$$

By Proposition 4.4, we may assume that for each integer $N \geq 1$ and each pair of integers $q > p \geq 0$,

$$\sum_{i=p}^{q-1} [f(x_i^N, x_{i+1}^N) - \mu(f)] \leq c(f) + 2. \tag{4.44}$$

Properties (c) and (g) and (4.28) imply that there exists an open set $V \subset \mathcal{M}$ in the relative topology for which

$$H_0^f \times H_0^f \subset V \text{ and } \bar{V} \subset V_f. \tag{4.45}$$

It follows from (ATP) that for each integer $N \geq 1$

$$\Omega(\{x_i^N\}_{i=0}^{\infty}) = H(f), \ d_1((x_i^N, x_{i+1}^N), H(f)) \to 0 \text{ as } i \to \infty. \tag{4.46}$$

By (4.46), (4.45), (4.28), and property (g) we may assume without loss of generality that for each pair of integers $N \geq 1$ and $i \geq 0$,

$$d_1((x_i^N, x_{i+1}^N), H(f)) \leq 4^{-1}\epsilon, \ (x_i^N, x_{i+1}^N) \in V. \tag{4.47}$$

By (4.43) for each integer $N \geq 1$ there exists an integer $T_N \geq 1$ such that

$$\text{dist}(H(f), \{(x_i^N, x_{i+1}^N) : i \in [T_N, T_N + N]\}) \geq \epsilon(7/8). \tag{4.48}$$

Together with (4.47) this implies that for each integer $N \geq 1$ there exists $h_N \in H(f)$ which satisfies

$$d_1(h_N, \{(x_i^N, x_{i+1}^N) : i \in [T_N, T_N + N]\}) \geq 2^{-1}\epsilon. \tag{4.49}$$

For each integer $N \geq 1$ we define a program $\{v_i^N\}_{i=0}^{\infty} \subset \mathcal{K}$ by

$$v_i^N = x_{i+T_N}^N, \ i = 0, 1, \ldots \tag{4.50}$$

It follows from (4.49) and (4.50) that

$$d_1(h_N, \{(v_i^N, v_{i+1}^N) : i = 0, \ldots, N\}) \geq 2^{-1}\epsilon \text{ for each integer } N \geq 1. \tag{4.51}$$

We may assume by extracting a subsequence and reindexing that

$$h_N \to h \in H(f) \text{ as } N \to \infty \tag{4.52}$$

and that there exists a sequence $\{v_i^*\}_{i=0}^{\infty} \subset \mathcal{K}$ such that

$$v_i^N \to v_i^* \text{ as } N \to \infty \text{ for each integer } i \geq 0. \tag{4.53}$$

Equations (4.51)–(4.53) imply that

$$d_1(h, (v_i^*, v_{i+1}^*)) \geq 4^{-1}\epsilon \text{ for each integer } i \geq 0. \tag{4.54}$$

On the other hand it follows from the lower semicontinuity of f, (4.44), (4.50), and (4.53) that $\{v_i^*\}_{i=0}^{\infty}$ is an (f)-good program. Combined with (ATP) and (4.52) this implies that

$$h \in H(f) = \Omega(\{v_i^*\}_{i=0}^{\infty}).$$

This contradicts (4.54). The obtained contradiction proves the lemma.

Lemma 4.14 *Let $\epsilon_0, M_0 > 0$ and let $l \geq 1$ be an integer such that for each (f)-good program $\{x_i\}_{i=0}^{\infty} \subset \mathcal{K}$,*

$$dist(H(f), \{(x_i, x_{i+1}) : i \in [p, p + l]\}) \leq 8^{-1}\epsilon_0 \tag{4.55}$$

for all large integers p (the existence of l follows from Lemma 4.13). Then there exists an integer $N \geq 10$ such that for each program $\{x_i\}_{i=0}^{Nl} \subset \mathcal{K}$ which satisfies

$$\sum_{i=0}^{Nl-1} f(x_i, x_{i+1}) \leq Nl\mu(f) + M_0 \tag{4.56}$$

there exists an integer $j_0 \in [0, N - 8]$ such that

$$dist(H(f), \{(x_i, x_{i+1}) : i \in [T, T + l]\}) \leq \epsilon_0 \tag{4.57}$$

for each integer $T \in [j_0 l, (j_0 + 7)l]$.

Proof Let us assume the contrary. Then for each integer $N \geq 10$ there exists a trajectory $\{x_i^N\}_{i=0}^{Nl} \subset \mathcal{K}$ such that

$$\sum_{i=0}^{Nl-1} f(x_i^N, x_{i+1}^N) \leq Nl\mu(f) + M_0 \tag{4.58}$$

and that for each integer $j \in [0, N - 8]$ there exists an integer $T(j) \in [jl, (j + 7)l]$ for which

$$dist(H(f), \{(x_i^N, x_{i+1}^N) : i \in [T(j), T(j) + l]\}) > \epsilon_0. \tag{4.59}$$

There exist a subsequence of programs $\{x_i^{N_k}\}_{i=0}^{N_k l}$, $k = 1, 2, \ldots$ and a sequence $\{y_i\}_{i=0}^{\infty} \subset \mathcal{K}$ such that

$$x_i^{N_k} \to y_i \text{ as } k \to \infty \text{ for each integer } i \geq 0. \tag{4.60}$$

It follows from (4.58) and property (b) that for each integer $N \geq 1$ and each pair of integers $q, p \in [0, N]$ satisfying $p < q$,

$$\sum_{i=p}^{q-1} f(x_i^N, x_{i+1}^N) \leq (q - p)\mu(f) + M_0 + 2c(f).$$

Together with (4.60) and lower semicontinuity of f this implies that $\{y_i\}_{i=0}^{\infty}$ is an (f)-good program. Therefore by the definition of l (see (4.55)) there exists an integer $Q \geq 1$ such that for each integer $T \geq Ql$,

$$\text{dist}(H(f), \{(y_i, y_{i+1}) : i \in [T, T+l]\}) \leq 8^{-1}\epsilon_0. \tag{4.61}$$

By (4.60) there exists an integer k such that

$$k \geq 3Q + 30, \ d(x_i^{N_k}, y_i) \leq 64^{-1}\epsilon_0, \ i = 0, \ldots, (2Q + 20)l. \tag{4.62}$$

It follows from (4.61) and (4.62) that for each integer $T \in [Ql, (Q+10)l]$,

$$\text{dist}(H(f), \{(x_i^{N_k}, x_{i+1}^{N_k}) : i \in [T, T+l]\}) \leq 4^{-1}\epsilon_0. \tag{4.63}$$

On the other hand it follows from the definition of $\{x_i^{N_k}\}_{i=0}^{N_k l}$ and (4.62) that there exists an integer $T(Q) \in [Ql, (Q+7)l]$ for which

$$\text{dist}(H(f), \{(x_i^{N_k}, x_{i+1}^{N_k}) : i \in [T(Q), T(Q)+l]\}) > \epsilon_0.$$

This contradicts (4.63) which holds for each integer $T \in [Ql, (Q+10)l]$. The obtained contradiction proves the lemma.

Lemma 4.14 and property (a) imply the following result.

Lemma 4.15 *Let $\epsilon_0, M_0 > 0$ and let $l \geq 1$ be an integer such that each (f)-good program $\{x_i\}_{i=0}^{\infty} \subset \mathcal{K}$ satisfies (4.55) for all sufficiently large integers p. Then there exist an integer $N \geq 10$ and a neighborhood $\mathcal{U}$ of f in $B(\mathcal{M})$ such that for each $g_i \in \mathcal{U}, \ i = 0, \ldots, Nl - 1$ and each program $\{x_i\}_{i=0}^{Nl} \subset \mathcal{K}$ which satisfies*

$$\sum_{i=0}^{Nl-1} g_i(x_i, x_{i+1}) \leq \inf\{U(\{g_i\}_{i=0}^{Nl-1}, y, z) : \ y, z \in \mathcal{K}\} + M_0$$

there exists an integer $j_0 \in [0, N - 8]$ such that (4.57) holds for each integer $T \in [j_0 l, (j_0 + 7)l]$.

Lemma 4.16 *Let $\epsilon_0, M_0 > 0$. Then there exist an integer $q_0 \geq 2$ and $\delta > 0$ such that for each integer $T \geq q_0$, each program $\{x_i\}_{i=0}^{T}$ which satisfies*

$$\sum_{i=0}^{T-1} f(x_i, x_{i+1}) \leq T\mu(f) + M_0 + \delta T \tag{4.64}$$

the following inequality holds:

$$\min\{d_1((x_i, x_{i+1}), H(f)) : \ i = 1, \ldots, T-1\} \leq \epsilon_0. \tag{4.65}$$

Proof By Lemma 4.14, there exist integers $l_0 \geq 1$ and $N_0 \geq 10$ such that for each program $\{y_i\}_{i=0}^{N_0 l_0} \subset \mathcal{K}$ satisfying

$$\sum_{i=0}^{N_0 l_0 - 1} f(y_i, y_{i+1}) \leq N_0 l_0 \mu(f) + 1 \tag{4.66}$$

there exists an integer $j_0 \in [0, N_0 - 8]$ such that

$$\text{dist}(H(f), \{(y_i, y_{i+1}) : i \in [\tau, \tau + l_0]\}) \leq \epsilon_0 \tag{4.67}$$

for each integer $\tau \in [j_0 l_0, (j_0 + 7)l_0]$.
    Choose a positive number

$$\delta < (2N_0 l_0)^{-1} \tag{4.68}$$

and a natural number

$$q_0 > (2N_0 l_0)[4 + M_0 + N_0 l_0(\|f\| + |\mu(f)|)]. \tag{4.69}$$

Assume that an integer $T \geq q_0$ and that a program $\{x_i\}_{i=0}^{T}$ satisfies (4.64). In order to complete the proof it is sufficient to show that (4.65) holds.
    Assume the contrary. Then

$$d_1((x_i, x_{i+1}), H(f)) > \epsilon_0, \ i = 1, \ldots, T - 1. \tag{4.70}$$

There is a natural number p such that

$$pN_0 l_0 \leq T < (p + 1)N_0 l_0. \tag{4.71}$$

In view of (4.69) and (4.71),

$$p \geq 4. \tag{4.72}$$

By (4.70) and the choice of l_0, N_0 (see (4.66), (4.67)), for each integer $j \in [0, p - 1]$,

$$\sum_{i=jN_0 l_0}^{(j+1)N_0 l_0 - 1} f(x_i, x_{i+1}) > N_0 l_0 \mu(f) + 1. \tag{4.73}$$

It follows from (4.64), (4.71), and (4.73) that

$$\delta T + M_0 \geq \sum_{i=0}^{T-1} (f(x_i, x_{i+1}) - \mu(f))$$

$$= \sum_{j=0}^{p-1} \left(\sum_{i=jN_0 l_0}^{(j+1)N_0 l_0 - 1} f(x_i, x_{i+1}) - \mu(f) \right)$$

$$+ \sum \{f(x_i, x_{i+1}) - \mu(f) : i \text{ is an integer and } pN_0 l_0 \leq i < T\}$$

$$\geq p - N_0 l_0(\|f\| + |\mu(f)|) \geq T(N_0 l_0)^{-1} - 1 - N_0 l_0(\|f\| + |\mu(f)|).$$

Together with (4.68) this implies that

$$q_0(2N_0 l_0)^{-1} \leq T(2N_0 l_0)^{-1} \leq T((N_0 l_0)^{-1} - \delta)$$

$$\leq 1 + M_0 + N_0 l_0(\|f\| + |\mu(f)|)$$

and

$$q_0 \leq (2N_0l_0)[1 + M_0 + N_0l_0(\|f\| + |\mu(f)|)].$$

This contradicts (4.69). The contradiction we have reached completes the proof of Lemma 4.16.

Lemma 4.17 *Let $\epsilon_0, M_0 > 0$. Then there exist an integer $Q \geq 1$ and a neighborhood $\mathcal{U}$ of f in $B(\mathcal{M})$ such that for each integer $q \geq 8Q + 8$, each $g_i \in \mathcal{U}$, $i = 0, \ldots, q - 1$, each program $\{x_i\}_{i=0}^q \subset \mathcal{K}$ which satisfies*

$$\sum_{i=0}^{q-1} g_i(x_i, x_{i+1}) \leq \inf\{U(\{g_i\}_{i=0}^{q-1}, y, z) : y, z \in \mathcal{K}\} + M_0 \tag{4.74}$$

and each integer $p \in [0, q - Q]$ the following inequality holds:

$$\min\{d_1((x_i, x_{i+1}), H(f)) : i \in [p, p + Q - 1]\} \leq \epsilon_0. \tag{4.75}$$

Proof Fix a number

$$\delta_0 \in (0, 8^{-1}\gamma(f)) \cap (0, 1) \tag{4.76}$$

(recall $\gamma(f)$ in properties (d)–(f)) such that

$$\{(x_1, x_2) \in \mathcal{M} : d(x_k, \omega(\{x_i^f\}_{i=0}^\infty)) \leq 8\delta_0, \ k = 1, 2\} \subset V_f. \tag{4.77}$$

By property (c) and (4.77) we may assume without loss of generality that

$$\epsilon_0 \in (0, 8^{-1}\delta_0) \tag{4.78}$$

and

$$|f(x_1, x_2) - f(x_3, x_4)| \leq 4^{-1} \tag{4.79}$$

for each $(x_1, x_2), (x_3, x_4) \in \mathcal{M}$ which satisfy

$$d(x_k, \omega(\{x_i^f\}_{i=0}^\infty)) \leq 8\delta_0, \ k = 1, 2, 3, 4, \ d_1((x_1, x_2), (x_3, x_4)) \leq 4\epsilon_0. \tag{4.80}$$

It follows from property (g) that there exist integers $N_1, N_2 \geq 1$ such that

$$\mathrm{dist}(\{(x_i^f, x_{i+1}^f) : i = 0, \ldots, N_1 - 1\}, H(f)) \leq 8^{-1}\epsilon_0,$$

$$\mathrm{dist}(\{(x_i^f, x_{i+1}^f) : i = 2N_1 + 4, \ldots, 2N_1 + 3 + N_2\}, H(f)) \leq 8^{-1}\epsilon_0. \tag{4.81}$$

By Lemma 4.16 there exists

$$\delta_1 \in (0, 16^{-1}) \tag{4.82}$$

and a natural number $Q_0 > 4$ such that the following property holds:

(Pi) for each integer $\tau \geq Q_0$ and each program $\{x_i\}_{i=0}^{\tau}$ which satisfies

$$\sum_{i=0}^{\tau-1} f(x_i, x_{i+1}) \leq \tau \mu(f) + 2\delta_1 \tau + M_0 + 4 + 3c(f) + 7(\|f\| + 1) + |\mu(f)|$$

we have

$$\min\{d_1((x_i, x_{i+1}), H(f)) : i = 1, \ldots, \tau - 1\} \leq \epsilon_0.$$

We may assume without loss of generality that

$$\delta_1 Q_0 < 1. \tag{4.83}$$

Fix an integer

$$Q > 4(Q_0 + N_1 + N_2 + 4). \tag{4.84}$$

Set

$$\mathcal{U} = \{g \in B(\mathcal{M}) : \|g - f\| < 2^{-1}\delta_1\}. \tag{4.85}$$

Assume that an integer $q \geq 8Q + 8$, $g_i \in \mathcal{U}$, $i = 0, \ldots, q - 1$ and that a program $\{x_i\}_{i=0}^{q} \subset \mathcal{K}$ satisfies (4.74). We show that for each pair of integers $k, s \in [0, q]$ satisfying

$$s - k \geq Q_0, \ k(q - s) = 0 \tag{4.86}$$

the following inequality holds:

$$\min\{d_1((x_i, x_{i+1}), H(f)) : i \in [k, s - 1]\} \leq \epsilon_0. \tag{4.87}$$

Let us assume the contrary. Then there exist integers $k, s \in [0, q]$ which satisfy (4.86) and such that

$$\min\{d_1((x_i, x_{i+1}), H(f)) : i \in [k, s - 1]\} > \epsilon_0. \tag{4.88}$$

We may assume without loss of generality that for each integer

$$p \in [0, q - 1] \cap \{k - 1, s\} \tag{4.89}$$

the following inequality holds:

$$d_1((x_p, x_{p+1}), H(f)) \leq \epsilon_0. \tag{4.90}$$

By (4.81) and (4.90) for any $p \in [0, q - 1] \cap \{k - 1, s\}$ there exists an integer $j(p) \in \{0, \ldots, N_1 - 1\}$ for which

$$d_1((x_p, x_{p+1}), (x_{j(p)}^f, x_{j(p)+1}^f)) \leq \epsilon_0(1 + 6^{-1}). \tag{4.91}$$

It follows from (4.91), (4.78), (4.76), and properties (d) and (g) that for any $p \in [0, q-1] \cap \{k-1, s\}$

$$\left(x_p, x^f_{j(p)+1}\right), \ \left(x^f_{j(p)}, x_{p+1}\right) \in \mathcal{M}. \tag{4.92}$$

By the definition of ϵ_0 (see (4.78)–(4.80)), (4.85), (4.92), and (4.91), for any $p \in [0, q-1] \cap \{k-1, s\}$ the following equation holds:

$$g_p\left(x^f_{j(p)}, x^f_{j(p)+1}\right), g_p\left(x_p, x^f_{j(p)+1}\right), \ g_p\left(x^f_{j(p)}, x_{p+1}\right)$$
$$\in [g_p(x_p, x_{p+1}) - 4, \ g_p(x_p, x_{p+1}) + 4]. \tag{4.93}$$

There are three cases: (i) $k = 0$, $s = q$; (ii) $k = 0$, $s < q$; (iii) $k > 0$, $s = q$. In all these cases we define a program $\{y_i\}_{i=0}^q \subset \mathcal{K}$ and estimate

$$\sum_{i=0}^{q-1} [f(y_i, y_{i+1}) - f(x_i, x_{i+1})].$$

In the case (i) we set

$$y_i = x^f_i, \ i = 0, \ldots, q. \tag{4.94}$$

In the case (ii) we set

$$y_i = x^f_{i+j(s)-s}, \ i = 0, \ldots, s, \ y_i = x_i, \ i = s+1, \ldots, q. \tag{4.95}$$

In the case (iii) we set

$$y_i = x_i, \ i = 0, \ldots, k-1, \ y_i = x^f_{i+j(k-1)-k+1}, \ i = k, \ldots, q. \tag{4.96}$$

By (4.92), the program $\{y_i\}_{i=0}^q \subset \mathcal{K}$ is well defined. It follows from (4.74), (4.93)–(4.96), (4.85), and the inclusion $g_i \in \mathcal{U}, i = 0, \ldots, q-1$ that

$$M_0 \geq \sum_{i=0}^{q-1} [g_i(x_i, x_{i+1}) - g_i(y_i, y_{i+1})] \geq \sum_{i=k}^{s-1} [g_i(x_i, x_{i+1}) - g_i(y_i.y_{i+1})] - 4$$

$$\geq \sum_{i=k}^{s-1} [f(x_i, x_{i+1}) - f(y_i, y_{i+1})] - (s-k)\delta_1 - 4. \tag{4.97}$$

Property (a) and (4.94)–(4.96) imply that

$$\sum_{i=k}^{s-1} f(y_i, y_{i+1}) \leq (s-k)\mu(f) + c(f). \tag{4.98}$$

By (4.97) and (4.98),

$$\sum_{i=k}^{s-1} f(x_i, x_{i+1}) \le M_0 + (s-k)\delta_1 + 4 + (s-k)\mu(f) + c(f).$$

It follows from the inequality above, (4.86) and property (Pi) that

$$\min\{d_1((x_i, x_{i+1}), H(f)) :\ i = k, \ldots, s-1\} \le \epsilon_0.$$

This contradicts (4.88). The obtained contradiction proves the following assertion:

(**B**) Eq. (4.87) holds for each pair of integers $k, s \in [0, q]$ which satisfy (4.86).

Assume that an integer $p \in [0, q - Q]$. We show that (4.75) holds. Let us assume the contrary. Then

$$d_1((x_i, x_{i+1}), H(f)) > \epsilon_0 \text{ for each integer } i \in [p, p + Q - 1]. \tag{4.99}$$

For each integer $j \in [0, q - 1]$ we set

$$D_j = \{i \in [0, j] \cap \mathbf{Z} :\ d_1((x_i, x_{i+1}), H(f)) \le \epsilon_0\},$$
$$C_j = \{i \in [j, q - 1] \cap \mathbf{Z} :$$
$$d_1((x_i, x_{i+1}), H(f)) \le \epsilon_0\}. \tag{4.100}$$

It follows from assertion (**B**), (4.99), (4.86), and (4.87) that

$$p > 0,\ D_{p-1} \ne \emptyset,\ p + Q < q,\ C_{p+Q} \ne \emptyset. \tag{4.101}$$

Set

$$k = \sup\{i :\ i \in D_{p-1}\},\ s = \inf\{i :\ i \in C_{p+Q}\}. \tag{4.102}$$

By (4.99)–(4.102)

$$0 \le k \le p - 1,\ p + Q \le s \le q - 1,\ d_1((x_i, x_{i+1}), H(f)) \le \epsilon_0,\ i = k, s,$$
$$d_1((x_i, x_{i+1}), H(f)) > \epsilon_0,\ i \in [k + 1, \ldots, s - 1]. \tag{4.103}$$

Set

$$m = \inf\{i :\ i \in D_k\},\ n = \sup\{i :\ i \in C_s\}. \tag{4.104}$$

Assertion (**B**), (4.104) and (4.100) imply that

$$m \le Q_0,\ n \ge q - Q_0. \tag{4.105}$$

It follows from (4.104) and (4.100) that

$$d_1((x_i, x_{i+1}), H(f)) \le \epsilon_0,\ i = m, n. \tag{4.106}$$

By (4.103), (4.106), and (4.81) there exist integers

$$i(k) \in [0, N_1 - 1], \ i(s) \in [2N_1 + 4, \ldots, 2N_1 + 3 + N_2],$$
$$i(n) \in [2N_1 + 4, \ldots, 2N_1 + 3 + N_2] \tag{4.107}$$

such that

$$d_1((x_l, x_{l+1}), (x^f_{i(l)}, x^f_{i(l)+1})) \le \epsilon_0 + 6^{-1}\epsilon_0, \ l = k, s, n. \tag{4.108}$$

It follows from (4.108), (4.78), (4.76), and properties (d) and (g) that

$$(x_k, x^f_{i(k)+1}) \in \mathcal{M}, \ (x^f_{i(s)}, x_{s+1}) \in \mathcal{M}, \ (x_n, x^f_{i(n)+1}) \in \mathcal{M}. \tag{4.109}$$

We define a program $\{y_i\}^q_{i=0}$ and estimate

$$\sum_{i=0}^{q-1} [g_i(x_i, x_{i+1}) - g_i(y_i, y_{i+1})].$$

Set

$$y_i = x_i, \ i = 0, \ldots, k, \ y_i = x^f_{i-k+i(k)}, \ i = k+1, \ldots, k+i(s) - i(k),$$
$$\tau = k + i(s) - i(k). \tag{4.110}$$

If $n = s$, then we set

$$y_i = x^f_{i-k+i(k)}, \ i = k + i(s) - i(k) + 1, \ldots, q. \tag{4.111}$$

Otherwise we put

$$y_i = x_{i+s-\tau}, \ i = k + i(s) - i(k) + 1, \ldots, n + k + i(s) - i(k) - s, \tag{4.112}$$
$$y_i = x^f_{i+i(n)-n-\tau+s}, \ i = n + k + i(s) - i(k) - s + 1, \ldots, q.$$

It follows from (4.109)–(4.112), (4.107), (4.103), and (4.104) that $\{y_i\}^q_{i=0}$ is a program. By (4.74), (4.85), (4.103), (4.110)–(4.112), (4.83), and the inclusion $g_i \in \mathcal{U}$, $i = 0, \ldots, q - 1$,

$$M_0 \ge \sum_{i=0}^{q-1} (g_i(x_i, x_{i+1}) - g_i(y_i, y_{i+1})) = \sum_{i=k}^{q-1} (g_i(x_i, x_{i+1}) - g_i(y_i, y_{i+1}))$$

$$\ge -\|f\| - 1 + \sum_{i=k+1}^{s-1} g_i(x_i, x_{i+1}) + \sum_{i=s}^{q-1} g_i(x_i, x_{i+1})$$

$$- (\|f\| + 1 + \sum_{i=k+1}^{q-1} g_i(y_i, y_{i+1})). \tag{4.113}$$

It follows from (4.85) and the inclusion $g_i \in \mathcal{U}, i = 0, \ldots, q - 1$ that

$$\sum_{i=k+1}^{s-1} g_i(x_i, x_{i+1}) \geq -\delta_1(s - k - 1) + \sum_{i=k+1}^{s-1} f(x_i, x_{i+1}). \qquad (4.114)$$

We estimate $\sum_{i=s}^{q-1} g_i(x_i, x_{i+1})$ and $\sum_{i=k+1}^{q-1} g_i(y_i, y_{i+1})$.

There are two cases: (1) $n = s$; (2) $n > s$. Consider the case (1). Then it follows from (4.85), (4.105), property (b), and the inclusion $g_i \in \mathcal{U}, i = 0, \ldots, q - 1$ that

$$\sum_{i=s}^{q-1} g_i(x_i, x_{i+1}) \geq -\delta_1 Q_0 + \sum_{i=s}^{q-1} f(x_i, x_{i+1}) \geq -\delta_1 Q_0 - c(f) + (q - s)\mu(f).$$

$$(4.115)$$

By (4.110), (4.111), (4.85), the inclusion $g_i \in \mathcal{U}, i = 0, \ldots, q - 1$, and properties (a) and (g),

$$\sum_{i=k+1}^{q-1} g_i(y_i, y_{i+1}) \leq \delta_1(q - k - 1) + \sum_{i=i(k)+1}^{q-k+i(k)-1} f(x_i^f, x_{i+1}^f)$$

$$\leq \delta_1(q - k - 1) + (q - k - 1)\mu(f) + c(f). \qquad (4.116)$$

Together with (4.115) this implies that

$$\sum_{i=s}^{q-1} g_i(x_i, x_{i+1}) - \sum_{i=k+1}^{q-1} g_i(y_i, y_{i+1}) \geq (k + 1 - s)\mu(f) - \delta_1 Q_0 - 2c(f)$$

$$- \delta_1(q - k - 1). \qquad (4.117)$$

Combining this equation with (4.113), (4.114), and the equality $n = s$ we obtain that

$$M_0 \geq -2\|f\| - 2 - \delta_1(s - k - 1) + \sum_{i=k+1}^{s-1} f(x_i, x_{i+1})$$

$$+\mu(f)(k + 1 - s) - \delta_1 Q_0 - 2c(f) - \delta_1(s - k - 1) - \delta_1(q - n). \qquad (4.118)$$

Consider the case (2). It follows from (4.110), (4.85), (4.107), (4.103), (4.105), (4.112), (4.83), the inclusion $g_i \in \mathcal{U}, i = 0, \ldots, q - 1$, and properties (a) and (g) that

$$\sum_{i=k+1}^{q-1} g_i(y_i, y_{i+1}) \leq \sum_{i=k+1}^{\tau-1} g_i(y_i, y_{i+1}) + \|f\| + 1 + \sum_{i=\tau+1}^{n-s+\tau} g_i(y_i, y_{i+1})$$

$$+ \sum_{i=n-s+\tau+1}^{q-1} g_i(y_i, y_{i+1})$$

$$\leq (i(s) - i(k))\delta_1 + \sum_{i=i(k)+1}^{i(s)-1} f(x_i^f, x_{i+1}^f) + \|f\| + 1 + \sum_{i=s+1}^{n} g_i(x_i, x_{i+1})$$

$$+2 + 2\|f\| + \sum_{i=i(n)+1}^{q+i(n)-n-\tau+s-1} f(x_i^f, x_{i+1}^f) + \delta_1(q - n - \tau + s - 1)$$

$$\leq \sum_{i=s+1}^{n} g_i(x_i, x_{i+1}) + 3\|f\|$$

$$+3 + \delta_1(q - n - k + s - 1) + \mu(f)(i(s) - i(k) - 1) + c(f)$$
$$+\mu(f)(q - n - \tau + s - 1) + c(f)$$

$$\leq \sum_{i=s+1}^{n} g_i(x_i, x_{i+1}) + 3\|f\| + 3 + 2c(f)$$

$$+\delta_1(q - n - k + s - 1) + \mu(f)(q - n - k + s - 2). \qquad (4.119)$$

By (4.85), the inclusion $g_i \in \mathcal{U}$, $i = 0, \ldots, q - 1$ and property (b)

$$\sum_{i=n}^{q-1} g_i(x_i, x_{i+1}) \geq -\delta_1(q - n) + \mu(f)(q - n) - c(f).$$

Together with (4.119), (4.114), (4.113), (4.85), and (4.105) this implies that

$$M_0 \geq -2\|f\| - 2 + \sum_{i=k+1}^{s-1} f(x_i, x_{i+1}) - \delta_1(s - k - 1)$$

$$+\sum_{i=s}^{n-1} g_i(x_i, x_{i+1}) + \sum_{i=n}^{q-1} g_i(x_i, x_{i+1}) - \sum_{i=k+1}^{q-1} g_i(y_i, y_{i+1})$$

$$\geq -2\|f\| - 2 + \sum_{i=k+1}^{s-1} f(x_i, x_{i+1}) - \delta_1(s - k - 1)$$

$$+\sum_{i=s}^{n-1} g_i(x_i, x_{i+1}) - \delta_1(q - n) + \mu(f)(q - n) - c(f)$$

$$-\sum_{i=s+1}^{n} g_i(x_i, x_{i+1}) - 3\|f\| - 3$$

$$-2c(f) - \delta_1(q - n - k + s - 1) - \mu(f)(q - n - k + s - 2)$$

$$\geq \sum_{i=k+1}^{s-1} f(x_i, x_{i+1}) - 7(\|f\| + 1)$$

$$-2\delta_1(s - k - 1) - 2\delta_1(q - n) - \mu(f)(s - k - 2) - 3c(f)$$

$$\geq \sum_{i=k+1}^{s-1} f(x_i, x_{i+1}) - 2\delta_1(s-k-1)$$

$$-\mu(f)(s-k-1) - 2\delta_1 Q_0 - 7(\|f\|+1) - 3c(f) - |\mu(f)|.$$

It follows from the relation above, (4.118), (4.105), and (4.83) that in both cases we have

$$\sum_{i=k+1}^{s-1} f(x_i, x_{i+1}) \leq \mu(f)(s-k-1)$$

$$+2\delta_1(s-k-1) + M_0 + 3c(f) + 7(\|f\|+1) + 2\delta_1 Q_0 + |\mu(f)|$$

$$\leq \mu(f)(s-k-1) + 2\delta_1(s-k-1) + M_0$$

$$+3c(f) + 7(\|f\|+1) + 2 + |\mu(f)|.$$

By the relation above, (4.84), (4.103), and property (Pi) there exists $i \in \{k+1, \ldots, s-1\}$ such that

$$d_1((x_i, x_{i+1}), H(f)) \leq \epsilon_0.$$

This contradicts (4.103). The obtained contradiction proves Lemma 4.17.

Lemma 4.18 *Let $M_0 > 0$. Then there exist numbers $M_1 > M_0$, $\epsilon > 0$, an integer $N_0 \geq 1$, and a neighborhood $\mathcal{U}$ of f in $B(\mathcal{M})$ such that for each each integer $q \geq 1$, each $g_i \in \mathcal{U}$, $i = 0, \ldots, q-1$, each program $\{x_i\}_{i=0}^q \subset \mathcal{K}$ which satisfies*

$$\sum_{i=0}^{q-1} g_i(x_i, x_{i+1}) \leq \inf\{U(\{g_i\}_{i=0}^{q-1}, y, z) : y, z \in \mathcal{K}\} + M_0 \qquad (4.120)$$

and each pair of integers $p_1, p_2 \in [0, q-1]$ which satisfy

$$p_2 > p_1 + N_0, \quad d_1((x_i, x_{i+1}), H(f)) \leq \epsilon, \quad i = p_1, p_2 \qquad (4.121)$$

the following inequality holds:

$$\sum_{i=p_1}^{p_2-1} g_i(x_i, x_{i+1}) \leq \inf\{U(\{g_i\}_{i=0}^{p_2-p_1-1}, y, z) : y, z \in \mathcal{K}\} + M_1. \qquad (4.122)$$

Proof By properties (c) and (d) there exist an open set $V \subset \mathcal{M}$ in the relative topology and a number $\delta_0 \in (0, 8^{-1})$ such that:

$$[\omega(\{x_i^f\}_{i=0}^\infty)]^2 \subset V \subset \bar{V} \subset V_f;$$

$$\{(x_1, x_2) \in \mathcal{M} : d(x_j, \omega(\{x_i^f\}_{i=0}^\infty)) \leq \delta_0, \ j = 1, 2\} \subset V; \qquad (4.123)$$

for each $(x_1, y_1) \in H(f)$ and each $(x_2, y_2) \in \mathcal{M}$ which satisfy

$$d_1((x_1, y_1), (x_2, y_2)) \leq \delta_0$$

we have

$$(x_1, y_2), \ (x_2, y_1) \in \mathcal{M}; \tag{4.124}$$

for each $(x_i, y_i) \in \bar{V}$, $i = 1, 2$ satisfying

$$d_1((x_1, y_1), (x_2, y_2)) \leq \delta_0 \tag{4.125}$$

the inequality $|f(x_1, y_1) - f(x_2, y_2)| \leq 1$ holds.

Fix

$$\epsilon \in (0, 64^{-1}\delta_0). \tag{4.126}$$

It follows from property (g) that there exist integers $N_1, N_2 \geq 1$ such that

$$\mathrm{dist}(\{(x_i^f, x_{i+1}^f) : \ i = 0, \ldots, N_1 - 1\}, H(f)) \leq 8^{-1}\epsilon,$$

$$\mathrm{dist}(\{(x_i^f, x_{i+1}^f) : \ i = 2N_1 + 4, \ldots, 2N_1 + 3 + N_2\}, \ H(f)) \leq 8^{-1}\epsilon. \tag{4.127}$$

By Lemma 4.17 there exist an integer $Q_1 \geq 1$ and a neighborhood $\mathcal{U}_1$ of f in $B(\mathcal{M})$ such that for each integer $q \geq 8Q_1 + 8$, each $g_i \in \mathcal{U}_1, i = 0, \ldots, q - 1$, each program $\{x_i\}_{i=0}^q \subset \mathcal{K}$ which satisfies

$$\sum_{i=0}^{q-1} g_i(x_i, x_{i+1}) \leq \inf\{U(\{g_i\}_{i=0}^{q-1}, y, z) : \ y, z \in \mathcal{K}\} + 4M_0 + 8 \tag{4.128}$$

and each integer $p \in [0, q - Q_1]$ the following equation holds:

$$\inf\{d_1((x_i, x_{i+1}), H(f)) : \ i \in [p, p + Q_1 - 1]\} \leq 4^{-1}\epsilon. \tag{4.129}$$

Fix integers N_3 and N_0 such that

$$N_3 \geq 10, \ N_0 \geq 64(Q_1 + N_1 + N_2 + N_3 + 4), \ 64^{-1}N_0 \in \mathbf{Z}, \tag{4.130}$$

a number

$$M_1 > M_0 + (1 + \|f\|)(N_1 + N_2 + N_0 + Q_1 + N_3 + 4)10 + 4 \tag{4.131}$$

and a neighborhood $\mathcal{U}$ of f in $B(\mathcal{M})$ such that

$$\mathcal{U} \subset \mathcal{U}_1 \cap \{g \in B(\mathcal{M}) : \ \|g - f\| < 1\}. \tag{4.132}$$

Assume that an integer $q \geq 1$, $g_i \in \mathcal{U}, i = 0, \ldots, q - 1$, a program $\{x_i\}_{i=0}^q \subset \mathcal{K}$ satisfies (4.120) and that integers $p_1, p_2 \in [0, q - 1]$ satisfy (4.121). We show that (4.122) holds.

Let us assume the contrary. Then

$$\sum_{i=p_1}^{p_2-1} g_i(x_i, x_{i+1}) > \inf\{U(\{g_i\}_{i=0}^{p_2-p_1-1}, y, z) : \ y, z \in \mathcal{K}\} + M_1. \tag{4.133}$$

There exists a program $\{z_i\}_{i=p_1}^{p_2} \subset \mathcal{K}$ such that

$$\sum_{i=p_1}^{p_2-1} g_i(z_i, z_{i+1})$$

$$< \min\{\sum_{i=p_1}^{p_2-1} g_i(x_i, x_{i+1}) - M_1, \inf\{U(\{g_i\}_{i=0}^{p_2-p_1-1}, y_1, y_2) : \ y_1, y_2 \in \mathcal{K}\} + 1\}. \tag{4.134}$$

It follows from (4.134) and the definition of Q_1 and $\mathcal{U}_1$ (see (4.128), (4.129)), (4.121), and (4.130) that there exist integers

$$j_1 \in [p_1 + Q_1 + 16^{-1}N_0, \ p_1 + 2Q_1 + 16^{-1}N_0],$$

$$j_2 \in [p_2 - 2Q_1 - 16^{-1}N_0, p_2 - Q_1 - 16^{-1}N_0] \tag{4.135}$$

such that

$$d_1((z_i, z_{i+1}), H(f)) \leq 2^{-1}\epsilon, \ i = j_1, j_2. \tag{4.136}$$

Set

$$p_3 = \sup\{i \in [p_2, q-1] : \ d_1((x_i, x_{i+1}), H(f)) \leq \epsilon\}. \tag{4.137}$$

By (4.121), (4.120), the definition of $Q_1, \mathcal{U}_1$ (see (4.128), (4.129)), and (4.130),

$$p_3 \geq p_2, \ p_3 \geq q - Q_1. \tag{4.138}$$

Equations (4.136), (4.121), (4.127), (4.138), and (4.137) imply that there exist integers

$$i_1, s_2 \in [0, N_1 - 1], \ i_3, s_1, i_2 \in [2N_1 + 4, 2N_1 + 3 + N_2] \tag{4.139}$$

such that

$$d_1((x_{i_k}^f, x_{i_k+1}^f), (x_{p_k}, x_{p_k+1})) \leq \epsilon + 6^{-1}\epsilon, \ k = 1, 2, 3, \tag{4.140}$$

$$d_1((x_{s_k}^f, x_{s_k+1}^f), (z_{j_k}, z_{j_k+1})) \leq \epsilon + 6^{-1}\epsilon, \ k = 1, 2.$$

We define a program $\{y_i\}_{i=0}^q$ and estimate

$$\sum_{i=0}^{q-1}[g_i(x_i,x_{i+1})-g_i(y_i,y_{i+1})].$$

Set

$$\tau(1)=p_1-i_1,\ \tau(2)=j_1-(s_1-i_1+p_1),\ \tau(3)=s_2-(j_2-j_1+s_1-i_1+p_1),$$
$$\tau(4)=p_2-(i_2-s_2+j_2-j_1+s_1-i_1+p_1),$$
$$\tau(5)=i_3-(p_3-p_2+i_2-s_2+j_2-j_1+s_1-i_1+p_1), \tag{4.141}$$

$$y_i=x_i,\ i=0,\ldots,p_1,\ y_i=x_{i-\tau(1)}^f,\ i=p_1+1,\ldots,s_1-i_1+p_1,$$

$$y_i=z_{i+\tau(2)},\ i=s_1-i_1+p_1+1,\ldots s_1-i_1+p_1+j_2-j_1,$$

$$y_i=x_{i+\tau(3)}^f,\ i=s_1-i_1+p_1+j_2-j_1+1,\ \ldots,i_2-s_2+j_2-j_1+s_1-i_1+p_1, \tag{4.142}$$

if $p_3-p_2\le N_3$, then we set

$$y_i=x_{i+\tau(3)}^f,\ i=i_2-s_2+j_2-j_1+s_1-i_1+p_1+1,\ldots,q, \tag{4.143}$$

otherwise we set

$$y_i=x_{i+\tau(4)},$$
$$i=i_2-s_2+j_2-j_1+s_1-i_1+p_1+1,\ \ldots,p_3-p_2+i_2-s_2+j_2-j_1+s_1-i_1+p_1,$$
$$y_i=x_{i+\tau(5)}^f,\ i=p_3-p_2+i_2-s_2+j_2-j_1+s_1-i_1+p_1+1,\ldots,q. \tag{4.144}$$

By (4.141)–(4.144), (4.139), (4.121), (4.130),(4.135), (4.140), (4.126), and the choice of δ_0 (see (4.124)), $\{y_i\}_{i=0}^q$ is a program. By (4.120) and (4.142),

$$M_0\ge\sum_{i=0}^{q-1}(g_i(x_i,x_{i+1})-g_i(y_i,y_{i+1}))$$

$$=\sum_{i=p_1}^{q-1}[g_i(x_i,x_{i+1})-g_i(y_i,y_{i+1})]. \tag{4.145}$$

By (4.141), (4.142), and (4.139),

$$\sum_{i=p_1}^{q-1}g_i(y_i,y_{i+1})\le 5\|f\|+5+\sum_{i=p_1+1}^{s_1+\tau(1)-1}g_i(y_i,y_{i+1})$$

$$+\sum_{i=s_1+\tau(1)+1}^{j_2-\tau(2)-1}g_i(y_i,y_{i+1})+\sum_{i=j_2-\tau(2)+1}^{q-1}g_i(y_i,y_{i+1})$$

$$\leq 5\|f\| + 5 + \sum_{i=i_1+1}^{s_1-1} g_i(x_i^f, x_{i+1}^f) + \sum_{i=j_1+1}^{j_2-1} g_i(z_i, z_{i+1}) + \sum_{i=j_2-\tau(2)+1}^{q-1} g_i(y_i, y_{i+1})$$

$$\leq (\|f\| + 1)(2N_1 + 8 + N_2) + \sum_{i=j_1+1}^{j_2-1} g_i(z_i, z_{i+1}) + \sum_{i=j_2-\tau(2)+1}^{q-1} g_i(y_i, y_{i+1}).$$

$$(4.146)$$

It follows from (4.134) and (4.135) that

$$\sum_{i=j_1+1}^{j_2-1} g_i(z_i, z_{i+1}) \leq \sum_{i=p_1}^{p_2-1} g_i(z_i, z_{i+1}) + 2(1 + \|f\|)(2Q_1 + 16^{-1}N_0 + 4)$$

$$\leq \sum_{i=p_1}^{p_2-1} g_i(x_i, x_{i+1}) - M_1 + 2(1 + \|f\|)(2Q_1 + 16^{-1}N_0 + 4). \quad (4.147)$$

There two cases: (i) $p_3 - p_2 \leq N_3$; (ii) $p_3 - p_2 > N_3$. Consider the case (i). Equations (4.135), (4.138), (4.139), (4.140), (4.141), and (4.142) imply that

$$q - (j_2 - \tau(2)) \leq q - p_2 + 4Q_1 + 8^{-1}N_0 \leq Q_1 + N_3 + 4Q_1 + 8^{-1}N_0.$$

Together with (4.146) and (4.148) this implies that

$$\sum_{i=p_1}^{q-1} g_i(y_i, y_{i+1}) \leq (1 + \|f\|)(2N_1 + N_2 + 8) + \sum_{i=p_1}^{p_2-1} g_i(x_i, x_{i+1}) - M_1$$

$$+ (1 + \|f\|)(2Q_1 + 16^{-1}N_0 + 4)2 + (1 + \|f\|)(5Q_1 + N_3 + 8^{-1}N_0)$$

$$\leq \sum_{i=p_1}^{p_2-1} g_i(x_i, x_{i+1}) - M_1 + (1 + \|f\|)(2N_1 + N_2 + 9Q_1 + N_3 + 4^{-1}N_0 + 16).$$

$$(4.148)$$

It follows from (4.138) that

$$\sum_{i=p_1}^{q-1} g_i(x_i, x_{i+1}) \geq \sum_{i=p_1}^{p_2-1} g_i(x_i, x_{i+1}) - (1 + \|f\|)(Q_1 + N_3).$$

Combining this equation with (4.145) and (4.148) we obtain that

$$M_0 \geq M_1 - (1 + \|f\|)[Q_1 + N_3 + 2N_1 + N_2 + 9Q_1 + N_3 + 4^{-1}N_0 + 16].$$

$$(4.149)$$

Consider the case (ii). By (4.141), (4.142), (4.144), (4.139), (4.138), (4.135), and the inclusion $g_i \in \mathcal{U}$, $i = 0, \ldots, q - 1$,

$$\sum_{i=j_2-\tau(2)+1}^{q-1} g_i(y_i, y_{i+1}) \leq (1 + \|f\|)(i_2 - s_2 + 2) + \sum_{i=p_2-\tau(4)+1}^{p_3-\tau(4)-1} g_i(y_i, y_{i+1})$$

$$+(1 + \|f\|)(q - (p_3 - p_2 + i_2 - s_2 + j_2 - j_1 + s_1 - i_1 + p_1) + 2)$$

$$\leq (1 + \|f\|)(2N_1 + N_2 + 5) + \sum_{i=p_2+1}^{p_3-1} g_i(x_i, x_{i+1})$$

$$+(1 + \|f\|)(Q_1 + 4Q_1 + 8^{-1}N_0 + 2)$$

$$\leq \sum_{i=p_2+1}^{p_3-1} g_i(x_i, x_{i+1}) + (1 + \|f\|)(2N_1 + N_2 + 5Q_1 + 8^{-1}N_0 + 7).$$

Together with (4.146) and (4.147) this implies that

$$\sum_{i=p_1}^{q-1} g_i(y_i, y_{i+1})$$

$$\leq (1 + \|f\|)(2N_1 + N_2 + 8) + \sum_{i=p_1}^{p_2-1} g_i(x_i, x_{i+1}) - M_1$$

$$+2(1 + \|f\|)(2Q_1 + 16^{-1}N_0 + 4)$$

$$+ \sum_{i=p_2+1}^{p_3-1} g_i(x_i, x_{i+1}) + (1 + \|f\|)(2N_1 + N_2 + 5Q_1 + 8^{-1}N_0 + 7)$$

$$\leq \sum_{i=p_1}^{p_3-1} g_i(x_i, x_{i+1}) - M_1 + (1 + \|f\|)(4N_1 + 2N_2 + 9Q_1 + 4^{-1}N_0 + 28).$$

Combining the equation above with (4.145) and (4.138) we obtain that

$$M_0 \geq -(Q_1 + 2)(1 + \|f\|) + M_1$$

$$-(1 + \|f\|)(4N_1 + 2N_2 + 4^{-1}N_0 + 9Q_1 + 28). \tag{4.150}$$

It both cases in view of (4.150), (4.149), and (4.132)

$$M_0 \geq M_1 - (1 + \|f\|)(4N_1 + 2N_2 + 4^{-1}N_0 + 10Q_1 + 2N_3 + 30).$$

This contradicts (4.131). The obtained contradiction proves Lemma 4.18.
Lemmas 4.17 and 4.18 imply the following result.

Lemma 4.19 *Let $M_0 > 0$. Then there exist a number $M_1 > 0$ and a neighborhood $\mathcal{U}$ of f in $B(\mathcal{M})$ such that for each integer $q \geq 1$, each $g_i \in \mathcal{U}$, $i = 0, \ldots, q - 1$, each program $\{x_i\}_{i=0}^q \subset \mathcal{K}$ which satisfies*

$$\sum_{i=0}^{q-1} g_i(x_i, x_{i+1}) \leq \inf\{U(\{g_i\}_{i=0}^{q-1}, y, z) : y, z \in \mathcal{K}\} + M_0 \tag{4.151}$$

and each pair of integers $p_1 \in [0, q-1]$, $p_2 \in (p_1, q]$ the following inequality holds:

$$\sum_{i=p_1}^{p_2-1} g_i(x_i, x_{i+1}) \leq \inf\{U(\{g_i\}_{i=0}^{p_2-p_1-1}, y, z) : y, z \in \mathcal{K}\} + M_1.$$

Lemmas 4.15 and 4.19 imply the following result.

Lemma 4.20 *Let $\epsilon_0, M_0 > 0$ and let $l \geq 1$ be an integer such that for each (f)-good program $\{x_i\}_{i=0}^{\infty} \subset \mathcal{K}$,*

$$dist(H(f), \{(x_i, x_{i+1}) : i \in [p, p+l]\}) \leq 8^{-1}\epsilon_0$$

for all sufficiently large natural numbers p (the existence of l follows from Lemma 4.13). Then there exist an integer $N \geq 10$ and a neighborhood $\mathcal{U}$ of f in $B(\mathcal{M})$ such that for each integer $q \geq 1$, each $g_i \in \mathcal{U}$, $i = 0, \ldots, q-1$, each program $\{x_i\}_{i=0}^{q} \subset \mathcal{K}$ which satisfies (4.151) and each integer p satisfying $0 \leq p \leq q - Nl$ there exists an integer $j_0 \in [0, N-8]$ such that for each integer $T \in [p + j_0 l, p + (j_0 + 7)l]$ the following inequality holds:

$$dist(H(f), \{(x_i, x_{i+1}) : i \in [T, T+l]\}) \leq \epsilon_0.$$

Lemma 4.21 *Let $\epsilon > 0$ and let $L \geq 1$ be an integer such that each (f)-good program $\{x_i\}_{i=0}^{\infty} \subset \mathcal{K}$ satisfies*

$$dist(H(f), \{(x_i, x_{i+1}) : i \in [p, p+L]\}) \leq \epsilon \tag{4.152}$$

for all sufficiently large integers $p \geq 0$ (the existence of L follows from Lemma 4.13). Then there exists a number $\delta > 0$ such that for each triplet of integers $T \geq L, s, q$ and each program $\{x_i\}_{i=0}^{T} \subset \mathcal{K}$ which satisfies

$$x_0 = x_s^f, \ x_T = x_q^f, \ \sum_{i=0}^{T-1} [f(x_i, x_{i+1}) - \mu(f)] - \pi^f(x_s^f) + \pi^f(x_q^f) \leq \delta$$

inequality (4.152) holds for all integers $p \in [0, T-L]$ (recall π^f in (4.29)).

Proof Assume the contrary. Then for each integer $k \geq 1$ there exist integers $T(k) \geq L, s(k), q(k), p(k)$, and a program $\{x_i^k\}_{i=0}^{T(k)} \subset \mathcal{K}$ such that

$$x_0^k = x_{s(k)}^f, \ x_{T(k)}^k = x_{q(k)}^f,$$

$$\sum_{i=0}^{T(k)-1} [f(x_i^k, x_{i+1}^k) - \mu(f)] - \pi^f(x_{s(k)}^f) + \pi^f(x_{q(k)}^f) \leq 2^{-k},$$

$$p(k) \in [0, T(k) - L],$$

$$dist(H(f), \{(x_i^k, x_{i+1}^k) : i \in [p(k), p(k) + L]\}) > \epsilon. \tag{4.153}$$

We construct an (f)-good program $\{y_i\}_{i=0}^{\infty}$.

Let $k \geq 1$ be an integer. By property (g)

$$(x_{s(k+1)-1}^f, x_{s(k+1)}^f) \in \Omega(\{x_i^f\}_{i=0}^{\infty}). \tag{4.154}$$

It follows from (4.154), properties (c) and (d) and Proposition 4.10 that there exists an integer $j(k)$ such that

$$j(k) > 2q(k) + 4, \ |\pi^f(x_{j(k)}^f) - \pi^f(x_{s(k+1)-1}^f)| \leq 2^{-k-1},$$

$$|\pi^f(x_{j(k)+1}^f) - \pi^f(x_{s(k+1)}^f)| \leq 2^{-k-1},$$

$$(x_{j(k)}^f, x_{s(k+1)}^f) \in \mathcal{M}, \ |f(x_{j(k)}^f, x_{s(k+1)}^f) - f(x_{s(k+1)-1}^f, x_{s(k+1)}^f)| \leq 2^{-k-1},$$

$$|f(x_{j(k)}^f, x_{j(k)+1}^f) - f(x_{s(k+1)-1}^f, x_{s(k+1)}^f)| \leq 2^{-k-1}. \tag{4.155}$$

Set

$$\tau(1) = T(1), \ \tau(k) = \sum_{i=1}^{k}(1 + T(i)) + \sum_{i=1}^{k-1}[j(i) - q(i)], \ k \in \mathbf{Z}_2.$$

We define

$$y_i = x_i^1, \ i = 0, \ldots, \tau(1), \ y_{\tau(k)+i} = x_{q(k)+i}^f, \ i = 1, \ldots, j(k) - q(k),$$

$$y_{\tau(k)+j(k)-q(k)+i} = x_{i-1}^{k+1}, \ i = 1 \ldots, 1 + T(k+1), \ k = 1, 2, 3, \ldots \tag{4.156}$$

It follows from (4.153), (4.155), (4.156), and Proposition 4.12 that

$$y_{\tau(k)} = x_{T(k)}^k, \ y_{\tau(k)+j(k)-q(k)} = x_{j(k)}^f, \ k = 1, 2, \ldots,$$

$\{y_i\}_{i=0}^{\infty}$ is a program and that for each integer $k \geq 1$,

$$\sum_{i=\tau(k)}^{\tau(k+1)-1} [f(y_i, y_{i+1}) - \mu(f)] - \pi^f(y_{\tau(k)}) + \pi^f(y_{\tau(k+1)})$$

$$= \sum_{i=\tau(k)}^{\tau(k)+j(k)-q(k)-1} [f(y_i, y_{i+1}) - \mu(f)] - \pi^f(y_{\tau(k)}) + \pi^f(y_{\tau(k)+j(k)-q(k)})$$

$$+[f(y_{\tau(k)+j(k)-q(k)}, y_{\tau(k)+j(k)-q(k)+1}) - \mu(f)$$

$$-\pi^f(y_{\tau(k)+j(k)-q(k)}) + \pi^f(y_{\tau(k)+j(k)-q(k)+1})]$$

$$+ \sum_{i=\tau(k)+j(k)-q(k)+1}^{\tau(k+1)-1} [f(y_i, y_{i+1}) - \mu(f)] - \pi^f(y_{\tau(k)+j(k)-q(k)+1}) + \pi^f(y_{\tau(k+1)})$$

$$= \sum_{i=q(k)}^{j(k)-1} [f(x_i^f, x_{i+1}^f) - \mu(f)] - \pi^f(x_{q(k)}^f) + \pi^f(x_{j(k)}^f)$$

$$+[f(x^f_{j(k)}, x^f_{s(k+1)}) - \mu(f) - \pi^f(x^f_{j(k)}) + \pi^f(x^f_{s(k+1)})]$$

$$+ \sum_{i=0}^{T(k+1)-1} [f(x_i^{k+1}, x_{i+1}^{k+1}) - \mu(f)] - \pi^f(x_0^{k+1}) + \pi^f(x_{T(k+1)}^{k+1})$$

$$\leq 2^{-k} + f(x^f_{s(k+1)-1}, x^f_{s(k+1)}) - \mu(f)$$

$$-\pi^f(x^f_{s(k+1)-1}) + \pi^f(x^f_{s(k+1)}) + 2^{-k-1} \leq 2^{-k+1}.$$

This implies that $\{y_i\}_{i=0}^{\infty}$ is an (f)-good program. Therefore, it follows from the choice of L that

$$\text{dist}(H(f), \{(y_i, y_{i+1}) : i \in [p, p + L]\}) \leq \epsilon \text{ for all large enough integers } p. \tag{4.157}$$

On the other hand by (4.153) and (4.156) for each integer $k \geq 1$,

$$\text{dist}(H(f), \{(y_i, y_{i+1}) : i \in [\tau(k) + j(k) - q(k) + 1 + p(k) + 1),$$

$$\tau(k) + j(k) - q(k) + 1 + p(k) + L]\}) > \epsilon.$$

This contradicts (4.157). The contradiction we have reached proves the lemma.

4.5   Proof of Theorem 4.5

There exist an (f)-minimal program $\{x_i^f\}_{i=-\infty}^{\infty}$, constants $c(f) > 0$, $\gamma(f) > 0$, $\mu(f) \in R^1$, an open set $V_f \subset \mathcal{M}$ in the relative topology such that properties (a)–(g) (see Sect. 4.4) hold. Note that we can use all the results of Sects. 4.3 and 4.4.

By Lemma 4.13 there exists an integer $L \geq 2$ such that for each (f)-good program $\{x_i\}_{i=0}^{\infty} \subset \mathcal{K}$ the equation

$$\text{dist}(H(f), \{(x_i, x_{i+1}) : i \in [s, s + L - 1]\}) \leq 4^{-1}\epsilon \tag{4.158}$$

holds for all large enough integers $s \geq 1$. By Lemma 4.21 there exists

$$\delta_0 \in (0, 8^{-1}\epsilon) \tag{4.159}$$

such that for each integer $T \geq L$ and each program $\{x_i\}_{i=0}^{T} \subset \mathcal{K}$ which satisfies

$$x_0, x_T \in \{x_i^f : i \in \mathbf{Z}\}, \quad \sum_{i=0}^{T-1} [f(x_i, x_{i+1}) - \mu(f)] - \pi^f(x_0) + \pi^f(x_T) \leq \delta_0 \tag{4.160}$$

the Eq. (4.158) holds for all integers $s \in [0, T - L]$.

Let an open set $V_f \subset \mathcal{M}$ in the relative topology be as guaranteed by property (c). By properties (c),(d), and (e) we may assume without loss of generality that

$$\{(x_1, x_2) \in \mathcal{M} : d(x_j, \omega(\{x_i^f\}_{i=0}^\infty)) \le 8\delta_0, \ j = 1, 2\} \subset V_f, \qquad (4.161)$$

for each $(x_1, y_1) \in H(f)$ and each $(x_2, y_2) \in \mathcal{M}$ satisfying

$$d_1((x_1, y_1), (x_2, y_2)) \le 8\delta_0$$

we have

$$(x_1, y_2), \ (x_2, y_1) \in \mathcal{M}, \qquad (4.162)$$

and that for each $(x_1, y_1) \in H(f)$ and each $(x_2, y_2) \in \mathcal{M}$ satisfying $d(y_1, x_2) \le 8\delta_0$ we have

$$(x_1, x_2) \in \mathcal{M}. \qquad (4.163)$$

Set

$$\|\pi^f\| = \sup\{|\pi^f(x)| : \ x \in H_0^f\}. \qquad (4.164)$$

Fix a natural number

$$Q > 24 + 2\delta_0^{-1}(M + 2 + 4(\|f\| + c(f) + |\mu(f)| + 1 + \epsilon + 2\|\pi^f\|)) \qquad (4.165)$$

(recall $c(f)$ in property (a)).

It follows from property (c), (4.161), and Proposition 4.10 that there exists

$$\delta_1 \in (0, 8^{-1}\delta_0) \qquad (4.166)$$

such that: for each

$$(y_1, y_2), \ (z_1, z_2) \in \mathcal{M} \cap \{x \in \mathcal{K} : \ d(x, \omega(\{x_i^f\}_{i=0}^\infty)) \le 8\delta_0\}^2 \qquad (4.167)$$

which satisfies

$$d(y_i, z_i) \le 8\delta_1, \ i = 1, 2 \qquad (4.168)$$

we have

$$|f(y_1, y_2) - f(z_1, z_2)| \le (16Q)^{-1}\delta_0; \qquad (4.169)$$

for each $y, z \in H_0^f$ which satisfy $d(y, z) \le 8\delta_1$ the equation

$$|\pi^f(y) - \pi^f(z)| \le (16Q)^{-1}\delta_0 \qquad (4.170)$$

holds. By property (g) there exist integers $N_1, N_2 \ge 1$ for which

$$\text{dist}(H(f), \{(x_i^f, x_{i+1}^f) : \ i = 0, \dots, N_1 - 1\}) \le 16^{-1}\delta_1, \qquad (4.171)$$

$$\text{dist}(H(f), \{(x_i^f, x_{i+1}^f) : i = 2N_1 + 4, \ldots, 2N_1 + 3 + N_2\}) \leq 16^{-1}\delta_1.$$

By Lemma 4.13 there exists an integer $L_1 \geq 1$ such that for each (f)-good program $\{x_i\}_{i=0}^{\infty} \subset \mathcal{K}$ the equation

$$\text{dist}(H(f), \{(x_i, x_{i+1}) : i \in [S, S + L_1]\}) \leq 8^{-1}\delta_1 \tag{4.172}$$

holds for all large enough natural numbers S.

We may assume that

$$L_1 \geq 8(L + N_1 + N_2 + 4). \tag{4.173}$$

By Lemma 4.20 there exist an integer $N \geq 10$ and a neighborhood $\mathcal{U}_1$ of f in $B(\mathcal{M})$ such that for each integer $P \geq 1$, each $g_i \in \mathcal{U}_1$, $i = 0, \ldots, P - 1$, each program $\{x_i\}_{i=0}^{P} \subset \mathcal{K}$ which satisfies

$$\sum_{i=0}^{P-1} g_i(x_i, x_{i+1}) \leq \inf\{U(\{g_i\}_{i=0}^{P-1}, y, z) : y, z \in \mathcal{K}\} + M + 4 \tag{4.174}$$

and each integer p satisfying $0 \leq p \leq P - NL_1$ there exists an integer $j_0 \in [0, N-8]$ such that for each integer $s \in [p + j_0 L_1, p + (j_0 + 7)L_1]$ the following equation holds:

$$\text{dist}(H(f), \{(x_i, x_{i+1}) : i \in [s, s + L_1]\}) \leq \delta_1. \tag{4.175}$$

Set

$$l = 50NL_1. \tag{4.176}$$

Fix a number

$$\delta_2 \in (0, (6400QN(L_1 + l))^{-1}\delta_1) \tag{4.177}$$

and a neighborhood $\mathcal{U}$ of f in $B(\mathcal{M})$ such that

$$\mathcal{U} \subset \mathcal{U}_1 \cap \{g \in B(\mathcal{M}) : \|g - f\| < \delta_2\}. \tag{4.178}$$

Assume that an integer $T \geq L + lQ$, $g_i \in \mathcal{U}$, $i = 0, \ldots, T - 1$ and that a program $\{x_i\}_{i=0}^{T} \subset \mathcal{K}$ satisfies

$$\sum_{i=0}^{T-1} g_i(x_i, x_{i+1}) \leq \inf\{U(\{g_i\}_{i=0}^{T-1}, y, z) : y, z \in \mathcal{K}\} + M. \tag{4.179}$$

Set

$$\mathcal{E} = \{s \in \mathbf{Z} : 10NL_1 \leq s \leq T - 10NL_1$$

$$\text{and dist}(H(f), \{(x_i, x_{i+1}) : i \in [s, s + L - 1]\}) > \epsilon\}. \tag{4.180}$$

If $\mathcal{E} = \emptyset$, then the assertion of the theorem is valid. Hence we may assume that $\mathcal{E} \neq \emptyset$. Set

$$h_1 = \inf\{h : h \in \mathcal{E}\}. \tag{4.181}$$

It follows from the choice of $\mathcal{U}_1$, N (see (4.174), (4.175)) and (4.179) that there are integers $i_1, i_2 \in \{0, \ldots, N - 8\}$ such that (4.175) holds for any integer

$$s \in [h_1 - (2N - i_1)L_1, h_1 - (2N - i_1 - 7)L_1] \cup [h_1 + (N + i_2)L_1, h_1$$
$$+ (N + i_2 + 7)L_1]. \tag{4.182}$$

Set

$$b_1 = h_1 - (2N - i_1)L_1, \quad c_1 = h_1 + (N + i_2)L_1. \tag{4.183}$$

By induction we define sequences of integers b_q, c_q, $q \geq 1$ such that
(B) $NL_1 \leq c_q - b_q \leq 4NL_1$, $b_q \geq c_{q-1}$ if $q \geq 2$,

$$[b_q, c_q - NL_1] \cap \mathcal{E} \neq \emptyset, \quad \mathcal{E} \setminus \cup_{j=1}^{q}[b_j, c_j] \subset (c_q, T];$$

(C) for $h \in \{b_q, c_q\}$ Eq. (4.175) holds for each $s \in [h, h + 7L_1]$.
It is easy to see that for $q = 1$ properties (B) and (C) hold.
Assume that sequences of integers $\{b_q\}_{q=1}^{k}$, $\{c_q\}_{q=1}^{k}$ have been defined and properties (B) and (C) hold for $q = 1, \ldots, k$ where k is a natural number. If

$$\mathcal{E} \setminus \cup_{q=1}^{k}[b_q, c_q] = \emptyset,$$

then the construction of the sequences is completed and b_k, c_k are their last elements.
Let us assume that

$$\mathcal{E} \setminus \cup_{q=1}^{k}[b_q, c_q] \neq \emptyset$$

and set

$$h_2 = \inf\{h : h \in \mathcal{E} \setminus \cup_{q=1}^{k}[b_q, c_q]\}.$$

It follows from (4.179), (4.180), and the definition of $\mathcal{U}_1$ (see (4.174), (4.175)) that there are integers $j_1, j_2 \in [0, N - 8]$ such that (4.175) holds for any integer

$$s \in [h_2 - (2N - j_1)L_1, h_2 - (2N - j_1 - 7)L_1] \cup [h_2 + (N + j_2)L_1, h_2 + (N + j_2 + 7)L_1].$$

Set

$$c_{k+1} = h_2 + (N + j_2)L_1, \quad b_{k+1} = \max\{c_k, h_2 - 2NL_1 + j_1L_1\}.$$

It is easy to see that properties (B) and (C) hold with $q = k + 1$. Evidently the construction of the sequence will be completed in a finite number of steps. Let $b_{\tilde{Q}}, c_{\tilde{Q}}$ be the last elements of the sequences. Clearly,

$$\mathcal{E} \subset \cup_{q=1}^{\tilde{Q}}[b_q, c_q]. \tag{4.184}$$

Let $i \in \{1, \dots, \tilde{Q}\}$. By property (C)

$$d_1((x_{b_i}, x_{b_i+1}), H(f)) \le \delta_1, \; d_1((x_{c_i}, x_{c_i+1}), H(f)) \le \delta_1.$$

It follows from these inequalities and the choice of N_1, N_2 (see (4.171)) that there exist

$$p(i) \in [0, N_1 - 1], \; s(i) \in [2N_1 + 4, \dots, 2N_1 + 3 + N_2] \qquad (4.185)$$

such that

$$d_1((x_{b_i}, x_{b_i+1}), (x^f_{p(i)}, x^f_{p(i)+1})) \le \delta_1 + 15^{-1}\delta_1,$$

$$d_1((x_{c_i}, x_{c_i+1}), (x^f_{s(i)}, x^f_{s(i)+1})) \le \delta_1 + 15^{-1}\delta_1. \qquad (4.186)$$

We show that

$$\sum_{j=b_i}^{c_i-1} [f(x_j, x_{j+1}) - \mu(f)] + \pi^f(x^f_{s(i)}) - \pi^f(x^f_{p(i)}) \ge \delta_0(1 - (8Q)^{-1}). \qquad (4.187)$$

Define a sequence $\{w_j\}_{j=b_i}^{c_i+1} \subset \mathcal{K}$ by

$$w_{b_i} = x^f_{p(i)}, \; w_j = x_j, \; j \in [b_i + 1, \dots, c_i], \; w_{c_i+1} = x^f_{s(i+1)}. \qquad (4.188)$$

By (4.188), (4.186), (4.166), (4.162), and (4.163) the sequence $\{w_j\}_{j=b_i}^{c_i+1}$ is a program.
  It follows from property (B), the choice of δ_0 (see (4.160)), (4.186), and (4.188) that

$$\sum_{j=b_i}^{c_i} [f(w_j, w_{j+1}) - \mu(f)] - \pi^f(w_{b_i}) + \pi^f(w_{c_i+1}) > \delta_0. \qquad (4.189)$$

By the choice of δ_1 (see (4.167)–(4.169)), (4.186), (4.188), and (4.166),

$$|f(w_j, w_{j+1}) - f(x_j, x_{j+1})| \le (16Q)^{-1}\delta_0, \; j = b_i, c_i,$$

$$|f(w_{c_i}, w_{c_i+1}) - f(x^f_{s(i)}, x^f_{s(i)+1})| \le (16Q)^{-1}\delta_0.$$

Together with (4.189) and (4.188) these inequalities imply that

$$\delta_0 < \pi^f(x^f_{s(i)+1}) - \pi^f(x^f_{p(i)})$$

$$+ \sum_{j=b_i}^{c_i-1} [f(x_j, x_{j+1}) - \mu(f)]$$

$$+ [f(w_{b_i}, w_{b_i+1}) - f(x_{b_i}, x_{b_i+1})] + f(w_{c_i}, w_{c_i+1}) - \mu(f)$$

$$\le \sum_{j=b_i}^{c_i-1} [f(x_j, x_{j+1}) - \mu(f)]$$

$$+(16Q)^{-1}\delta_0 + (16Q)^{-1}\delta_0 + f(x_{s(i)}^f, x_{s(i)+1}^f) - \mu(f)$$
$$+\pi^f(x_{s(i)+1}^f) - \pi^f(x_{p(i)}^f).$$

Equation (4.187) follows from the equation above. By Proposition 4.12, (4.187), (4.178), and property (B),

$$\sum_{j=b_i}^{c_i-1} [g_j(x_j, x_{j+1}) - \mu(f)] + \pi^f(x_{s(i)}^f) - \pi^f(x_{p(i)}^f)$$

$$\geq \delta_0(1 - (8Q)^{-1}) - \delta_2 4NL_1. \tag{4.190}$$

Define a sequence $\{u_j^i\}_{j=0}^{s(i)-p(i)} \subset \mathcal{K}$ by

$$u_0^i = x_{b_i}, \ u_j^i = x_{p(i)+j}^f, \ j = 1, \ldots, s(i) - p(i) - 1, \ u_{s(i)-p(i)}^i = x_{c_i}. \tag{4.191}$$

It follows from (4.191), (4.186), (4.166), (4.162), and (4.163) that the sequence $\{u_j^i\}_{j=0}^{s(i)-p(i)}$ is a program. By (4.191), Proposition 4.12, (4.186), and (4.166)–(4.169)

$$\sum_{j=0}^{s(i)-p(i)-1} [f(u_j^i, u_{j+1}^i) - \mu(f)] - \pi^f(x_{p(i)}^f) + \pi^f(x_{s(i)}^f)$$

$$= f(x_{b_i}, x_{p(i)+1}^f) - f(x_{p(i)}^f, x_{p(i)+1}^f) + f(x_{s(i)-1}^f, x_{c_i}) - f(x_{s(i)-1}^f, x_{s(i)}^f)$$

$$+ \sum_{j=p(i)}^{s(i)-1} [f(x_j^f, x_{j+1}^f) - \mu(f)] - \pi^f(x_{p(i)}^f) + \pi^f(x_{s(i)}^f) \leq (8Q)^{-1}\delta_0.$$

This equation, (4.178), (4.185), and (4.173) imply that

$$\sum_{j=0}^{s(i)-p(i)-1} [g_i(u_j^i, u_{j+1}^i) - \mu(f)] - \pi^f(x_{p(i)}^f) + \pi^f(x_{s(i)}^f) \leq (8Q)^{-1}\delta_0 + \delta_2 L_1.$$

$$\tag{4.192}$$

By induction we define sequences of integers $\bar{b}(i), \bar{c}(i), i = 1, \ldots, \tilde{Q}$ by

$$\bar{b}(1) = b_1, \ \bar{c}(i) = \bar{b}(i) + s(i) - p(i), \ i = 1, \ldots, \tilde{Q},$$

$$\bar{b}(i+1) = b_{i+1} + \bar{c}(i) - c_i \text{ for all integers } i \text{ such that } i \leq \tilde{Q} - 1. \tag{4.193}$$

By (4.193), property (B), (4.185), and (4.173),

$$c_{\tilde{Q}} - \bar{c}(\tilde{Q}) = \sum_{i=1}^{\tilde{Q}} [c_i - b_i - (s(i) - p(i))] \in [\tilde{Q} L_1 (N-1), 4\tilde{Q} L_1 N]. \tag{4.194}$$

Define a sequence $\{y_j\}_{j=0}^{c_{\tilde{Q}}} \subset \mathcal{K}$ by

$$y_j = x_j, \ j = 0, \ldots, b_1, \ y_j = u_{j-\bar{b}(i)}^i, \ j \in [\bar{b}(i), \bar{c}(i)], \ i = 1, \ldots, \tilde{Q},$$

$$y_j = x_{j-\bar{c}(i)+c_i}, \ j \in [\bar{c}(i), \bar{b}(i+1)] \text{ for all natural numbers } i \le \tilde{Q} - 1,$$

$$y_j = x^f_{j-\bar{c}(\tilde{Q})+s(\tilde{Q})}, \ j = \bar{c}(\tilde{Q}) + 1, \ldots, c_{\tilde{Q}}. \tag{4.195}$$

It follows from (4.195), (4.193), (4.191), (4.186), (4.162), and (4.163) that $\{y_j\}_{j=0}^{c_{\tilde{Q}}}$ is well defined and is a program. Evidently,

$$y_{c_{\tilde{Q}}} = x^f_{c_{\tilde{Q}}-\bar{c}(\tilde{Q})+s(\tilde{Q})}. \tag{4.196}$$

Equation (4.171) implies that there exists an integer $q \in [0, \ldots, N_1 - 1]$ for which

$$d_1((x^f_{c(\tilde{Q})-\bar{c}(\tilde{Q})+s(\tilde{Q})}, x^f_{c(\tilde{Q})-\bar{c}(\tilde{Q})+s(\tilde{Q})+1}), (x^f_q, x^f_{q+1})) \le 15^{-1}\delta_1. \tag{4.197}$$

Together with (4.196), (4.162), (4.163), and (4.166) this implies that

$$(y_{c_{\tilde{Q}}}, x^f_{q+1}) \in \mathcal{M}. \tag{4.198}$$

By (4.186), which holds with $i = \tilde{Q}$, and (4.162), (4.163), and (4.166)

$$(x^f_{s(\tilde{Q})}, x_{c_{\tilde{Q}}+1}) \in \mathcal{M}. \tag{4.199}$$

We set

$$y_{c_{\tilde{Q}}+j} = x^f_{q+s}, \ j = 1, \ldots, s(\tilde{Q}) - q,$$

$$y_{c_{\tilde{Q}}+s(\tilde{Q})-q+j} = x_{c_{\tilde{Q}}+j}, \ j = 1, \ldots, T - c_{\tilde{Q}} - s(\tilde{Q}) + q. \tag{4.200}$$

Equations (4.200), (4.198), and (4.199) imply that $\{y_j\}_{j=0}^T \subset \mathcal{K}$ is a program. It follows from (4.179), (4.195), (4.193), (4.190), and (4.192) that

$$M \ge \sum_{i=0}^{T-1} [g_i(x_i, x_{i+1}) - \mu(f)] - \sum_{i=0}^{T-1} [g_i(y_i, y_{i+1}) - \mu(f)]$$

$$= \sum_{j=1}^{\tilde{Q}} \sum_{i=b_j}^{c_j-1} [g_i(x_i, x_{i+1}) - \mu(f)] + \sum_{j \in \mathbf{Z}: \, 1 \le j \le \tilde{Q}-1} \sum_{i=c_j}^{b_{j+1}-1} [g_i(x_i, x_{i+1}) - \mu(f)]$$

$$+ \sum_{i=c_{\tilde{Q}}}^{T-1} [g_i(x_i, x_{i+1}) - \mu(f)] - \sum_{j=1}^{\tilde{Q}} \sum_{i=\bar{b}(j)}^{\bar{c}(j)-1} [g_i(y_i, y_{i+1}) - \mu(f)]$$

$$- \sum_{j \in \mathbf{Z}: \, 1 \le j \le \tilde{Q}-1} \sum_{i=\bar{c}(j)}^{\bar{b}(j+1)-1} [g_i(y_i, y_{i+1}) - \mu(f)] - \sum_{i=\bar{c}_{\tilde{Q}}}^{T-1} [g_i(y_i, y_{i+1}) - \mu(f)]$$

$$\ge \tilde{Q}(\delta_0(1 - (8Q)^{-1}) - \delta_2 4NL_1 - \delta_0(8Q)^{-1} - \delta_2 L_1)$$

$$+ \sum_{i=c_{\tilde{Q}}}^{T-1} [g_i(x_i, x_{i+1}) - \mu(f)] - \sum_{i=\bar{c}(\tilde{Q})}^{T-1} [g_i(y_i, y_{i+1}) - \mu(f)]. \tag{4.201}$$

We estimate $\sum_{i=\bar{c}(\tilde{Q})}^{T-1} [g_i(y_i, y_{i+1}) - \mu(f)]$. In view of (4.195), (4.191), (4.200), and (4.178)

$$\sum_{i=\bar{c}(\tilde{Q})}^{T-1} [g_i(y_i, y_{i+1}) - \mu(f)] \leq 1 + \|f\| + |\mu(f)| + \sum_{i=s(\tilde{Q})+1}^{c_{\tilde{Q}} - \bar{c}(\tilde{Q}) + s(\tilde{Q}) - 1} [g_i(x_i^f, x_{i+1}^f) - \mu(f)]$$

$$+ 1 + \|f\| + |\mu(f)| + \sum_{i=q+1}^{s(\tilde{Q})-1} [g_i(x_i^f, x_{i+1}^f) - \mu(f)] + 1 + \|f\| + |\mu(f)|$$

$$+ \sum_{i=c_Q+1}^{T-s(\tilde{Q})+q-1} [g_i(x_i, x_{i+1}) - \mu(f)] \leq 4(1 + \|f\| + |\mu(f)|)$$

$$+ \delta_2(c_{\tilde{Q}} - \bar{c}(\tilde{Q})) + \sum_{i=s(\tilde{Q})+1}^{c_{\tilde{Q}} - \bar{c}(\tilde{Q}) + s(\tilde{Q}) - 1} [f(x_i^f, x_{i+1}^f) - \mu(f)] + \delta_2 s(\tilde{Q})$$

$$+ \sum_{i=q+1}^{s(\tilde{Q})-1} [f(x_i^f, x_{i+1}^f) - \mu(f)]$$

$$+ \sum_{i=c_{\tilde{Q}}}^{T-1} [g_i(x_i, x_{i+1}) - \mu(f)] + \delta_2(s(\tilde{Q}) + 1) - \sum_{i=T-s(\tilde{Q})-q}^{T-1} [f(x_i, x_{i+1}) - \mu(f)].$$

It follows from this equation, (4.194), (4.185), (4.173), and properties (a) and (b) that

$$\sum_{i=\bar{c}(\tilde{Q})}^{T-1} [g_i(y_i, y_{i+1}) - \mu(f)]$$

$$\leq \sum_{i=c_{\tilde{Q}}}^{T-1} [g_i(x_i, x_{i+1}) - \mu(f)] + 4(1 + \|f\| + |\mu(f)|) + \delta_2 8 \tilde{Q} L_1 N + 3c(f).$$

Together with (4.178), (4.201), (4.177), and (4.166) this implies that

$$M \geq \tilde{Q}[\delta_0(1 - (4Q)^{-1}) - \delta_2 8 N L_1] - (1 + \|f\| + |\mu(f)|)4 - \delta_2 8 \tilde{Q} L_1 N - 3c(f)$$

$$\geq 2^{-1} \tilde{Q} \delta_0 - (1 + \|f\| + |\mu(f)|)4 - 3c(f).$$

By this equation and (4.165), $\tilde{Q} \leq Q - 24$. This completes the proof of Theorem 4.5.

References

1. Anderson, B.D.O., Moore, J.B.: Linear Optimal Control. Prentice-Hall, Englewood Cliffs (1971)
2. Arkin, V.I., Evstigneev, I.V.: Stochastic Models of Control and Economic Dynamics. Academic Press, London (1987)
3. Aseev, S.M., Kryazhimskiy, A.V.: The Pontryagin maximum principle and transversality conditions for a class of optimal control problems with infinite time horizons. SIAM J. Control Optim. **43**(3), 1094–1119 (2004)
4. Aseev, S.M., Veliov, V.M.: Maximum principle for infinite-horizon optimal control problems with dominating discount. Dyn. Cont. Discrete Impuls. Syst. Ser. B **19**(1–2), 43–63 (2012)
5. Atsumi, H.: Neoclassical growth and the efficient program of capital accumulation. Rev. Econ. Stud. **32**(2), 127–136 (1965)
6. Aubry, S., Le Daeron, P.Y.: The discrete Frenkel–Kontorova model and its extensions I. Phys. D: Nonlinear Phenom. **8**, 381–422 (1983)
7. Baumeister, J., Leitao, A., Silva, G.N.: On the value function for nonautonomous optimal control problem with infinite horizon. Syst. Control Lett. **56**(3), 188–196 (2007)
8. Blot, J.: Infinite-horizon Pontryagin principles without invertibility. J. Nonlinear Convex Anal. **10**(2), 177–189 (2009)
9. Blot, J., Cartigny, P.: Optimality in infinite-horizon variational problems under sign conditions. J. Optim. Theory Appl. **106**(2), 411–419 (2000)
10. Blot, J., Hayek, N.: Sufficient conditions for infinite-horizon calculus of variations problems. ESAIM Control Optim. Calc. Var. **5**, 279–292 (2000)
11. Bright, I.: A reduction of topological infinite-horizon optimization to periodic optimization in a class of compact 2-manifolds. J. Math. Anal. Appl. **394**(1), 84–101 (2012)
12. Carlson, D.A.: The existence of catching-up optimal solutions for a class of infinite horizon optimal control problems with time delay. SIAM J. Control Optim. **28**(2), 402–422 (1990)
13. Carlson, D.A., Haurie, A., Leizarowitz, A.: Infinite Horizon Optimal Control. Springer-Verlag, Berlin (1991)
14. Cartigny, P., Michel, P.: On a sufficient transversality condition for infinite horizon optimal control problems. Automatica **39**(6), 1007–1010 (2003)
15. Coleman, B.D., Marcus, M., Mizel, V.J.: On the thermodynamics of periodic phases. Arch. Rational Mech. Anal. **117**(4), 321–347 (1992)
16. Gaitsgory, V., Rossomakhine, S., Thatcher, N.: Approximate solution of the HJB inequality related to the infinite horizon optimal control problem with discounting. Dyn. Cont. Discrete Impuls. Syst. Ser. B **19**, 65–92 (2012)
17. Gale, D.: On optimal development in a multi-sector economy. Rev. Econ. Stud. **34**, 1–18 (1967)
18. Guo, X., Hernandez-Lerma, O.: Zero-sum continuous-time Markov games with unbounded transition and discounted payoff rates. Bernoulli **11**, 1009–1029 (2005)

A. J. Zaslavski, *Stability of the Turnpike Phenomenon in Discrete-Time Optimal Control Problems*, SpringerBriefs in Optimization, DOI 10.1007/978-3-319-08034-5, © The Author 2014

19. Hayek, N.: Infinite horizon multiobjective optimal control problems in the discrete time case. Optimization **60**, 509–529 (2011)
20. Jasso-Fuentes, H., Hernandez-Lerma, O.: Characterizations of overtaking optimality for controlled diffusion processes. Appl. Math. Optim. **57**, 349–369 (2008)
21. Kolokoltsov, V., Yang, W.: The turnpike theorems for Markov games. Dyn. Games Appl. **2**(3), 294–312 (2012)
22. Leizarowitz, A.: Infinite horizon autonomous systems with unbounded cost. Appl. Math. Optim. **13**(1), 19–43 (1985)
23. Leizarowitz, A.: Tracking nonperiodic trajectories with the overtaking criterion. Appl. Math. Optim. **14**, 155–171 (1986)
24. Leizarowitz, A., Mizel, V.J.: One dimensional infinite horizon variational problems arising in continuum mechanics. Arch. Ration. Mech. Anal. **106**(2), 161–194 (1989)
25. Lykina, V., Pickenhain, S., Wagner, M.: Different interpretations of the improper integral objective in an infinite horizon control problem. J. Math. Anal. Appl. **340**, 498–510 (2008)
26. Makarov, V.L., Rubinov, A.M.: Mathematical Theory of Economic Dynamics and equilibria. Springer-Verlag, New York (1977)
27. Malinowska, A.B., Martins, N., Torres, D.F.M.: Transversality conditions for infinite horizon variational problems on time scales. Optim. Lett. **5**, 41–53 (2011)
28. Marcus, M., Zaslavski, A.J.: On a class of second order variational problems with constraints. Israel J. Math. **111**, 1–28 (1999)
29. Marcus, M., Zaslavski, A.J.: The structure of extremals of a class of second order variational problems. Ann. Inst. H. Poincaré, Anal. non linéaire Anal. **16**(5), 593–629 (1999)
30. Marcus, M., Zaslavski, A.J.: The structure and limiting behavior of locally optimal minimizers. Ann. Inst. H. Poincaré, Anal. non linéaire Anal. **19**(3), 343–370 (2002)
31. McKenzie, L.W.: Turnpike theory. Econometrica **44**, 841–866 (1976)
32. Mordukhovich, B.S.: Minimax design for a class of distributed parameter systems. Automat. Remote Control **50**, 1333–1340 (1990)
33. Mordukhovich, B.S.: Optimal control and feedback design of state-constrained parabolic systems in uncertainly conditions. Appl. Anal. **90**, 1075–1109 (2011)
34. Mordukhovich, B.S., Shvartsman, I.: Optimization and feedback control of constrained parabolic systems under uncertain perturbations. Optimal Control, Stabilization and Non-smooth Analysis. Lecture notes control information science, vol. 301 Springer, 121–132 (2004)
35. Ocana Anaya, E., Cartigny, P., Loisel, P.: Singular infinite horizon calculus of variations. Applications to fisheries management. J. Nonlinear Convex Anal. **10**(2), 157–176 (2009)
36. Pickenhain, S., Lykina, V., Wagner, M.: On the lower semicontinuity of functionals involving Lebesgue or improper Riemann integrals in infinite horizon optimal control problems. Control Cybernet. **37**, 451–468 (2008)
37. Rockafellar, R.T.: Convex Analysis. Princeton University Press, Princeton (1970)
38. Rubinov, A.M.: Economic dynamics. J. Soviet Math. **26**, 1975–2012 (1984)
39. Samuelson, P.A.: A catenary turnpike theorem involving consumption and the golden rule. Amer. Econom. Rev. **55**, 486–496 (1965)
40. von Weizsacker, C.C.: Existence of optimal programs of accumulation for an infinite horizon. Rev. Econ. Stud. **32**, 85–104 (1965)
41. Zaslavski, A.J.: Ground states in Frenkel–Kontorova model. Math. USSR Izvestiya **29**, 323–354 (1987)
42. Zaslavski, A.J.: Optimal programs on infinite horizon 1. SIAM J. Control Optim. **33**(6), 1643–1660 (1995)
43. Zaslavski, A.J.: Optimal programs on infinite horizon 2. SIAM J. Control and Optim. **33**(6), 1661–1686 (1995)
44. Zaslavski, A.J.: Turnpike property for dynamic discrete time zero-sum games. Abstr. Appl. Anal. **4**(1), 21–48 (1999)

45. Zaslavski, A.J.: Turnpike Properties in the Calculus of Variations and Optimal Control. Springer, New York (2006)
46. Zaslavski, A.J.: Turnpike results for a discrete-time optimal control systems arising in economic dynamics. Nonlinear Anal. Theory, Meth. Appl. **67**(7), 2024–2049 (2007)
47. Zaslavski, A.J.: Two turnpike results for a discrete-time optimal control systems. Nonlinear Anal. **71**(12), 902–909 (2009)
48. Zaslavski, A.J.: Stability of a turnpike phenomenon for a discrete-time optimal control systems. J. Optim. Theory Appl. **145**(3), 597–612 (2010)
49. Zaslavski, A.J.: Turnpike properties of approximate solutions for discrete-time control systems. Commun. Math. Anal. **11**, 36–45 (2011)
50. Zaslavski, A.J.: Structure of approximate solutions for a class of optimal control systems. J. Math. Appl. **34**, 1–14 (2011)
51. Zaslavski, A.J.: Stability of a turnpike phenomenon for a class of optimal control systems in metric spaces. Numerical Algebra Control Optim. **1**, 245–260 (2011)
52. Zaslavski, A.J.: The existence and structure of approximate solutions of dynamic discrete time zero-sum games. J. Nonlinear Convex Anal. **12**, 49–68 (2011)
53. Zaslavski, A.J.: A generic turnpike result for a class of discrete-time optimal control systems. Dyn. Contin. Discrete Impuls. Syst. Ser. B **19**, 225–265 (2012)
54. Zaslavski, A.J.: Existence and structure of solutions for a class of optimal control systems with discounting arising in economic dynamics. Nonlinear Anal. Real World Appl. **13**(4), 1749–1760 (2012)
55. Zaslavski, A.J.: Structure of Solutions of Variational Problems. SpringerBriefs in Optimization, New York (2013)
56. Zaslavski, A.J.: Structure of Approximate Solutions of Optimal Control Problems. SpringerBriefs in Optimization, New York (2013)
57. Zaslavski, A.J., Leizarowitz, A.: Optimal solutions of linear control systems with nonperiodic integrands. Math. Op. Res. **22**(3), 726–746 (1997)

Index

A. J. Zaslavski, *Stability of the Turnpike Phenomenon in Discrete-Time*
Optimal Control Problems, SpringerBriefs in Optimization,
DOI 10.1007/978-3-319-08034-5, © The Author 2014